Spotlight SCIENCE

9

Keith JOHNSON ★ **Sue ADAMSON** ★ **Gareth WILLIAMS** ★ **Lawrie RYAN**

With the active support of: Bob Wakefield, Roger Frost, Helen Davis, Phil Bunyan, Michael Cotter, Cathryn Mellor, Peter Borrows, John Bailey, Janet Hawkins, Sarah Ryan, Judy Ryan, Ann Johnson, Graham Adamson, Diana Williams.

FRAMEWORK EDITION

First published in 2004 by:
Nelson Thornes Ltd
Delta Place
27 Bath Road
CHELTENHAM
GL53 7TH
United Kingdom

04 05 06 07 08 / 10 9 8 7 6 5 4 3 2 1

A catalogue record for this book is available from the British Library

ISBN 0 7487 7474 2

Illustrations by Jane Cope, Angela Lumley, Mike Gordon, Bill Piggins and Peters & Zabransky
Page make-up by Tech-Set

Printed and bound in China by Midas

Acknowledgements

The authors and publishers are grateful to the following for permission to reproduce photographs:

Ace Picture Library: Roger Howard 96T; Action Plus: Glyn Kirk 128T, 148T; Adams Picture Library: 70M; AKG London: 106B; Alamy: Brand X Pictures 81B, David Hoffman 91, Worldwide Picture Library 95T, Phototake Inc. 131; ALCAN: 103M; Allsport: Gray Mortimore 146T; Ann Ronan Picture Library: 36M, 137T; Associated Sports Photos: 18c, George Herringshaw 18g; Axon Images: 57L, M, 83 (label design); Barry Gorman: 152; Biophoto Associates: 58B; BMW: 95B; Bob Martin: 33B; British Steel: 100B,T; Bruce Coleman: Hans Reinhard 8BL; British Rail: 74; BT Pictures: 134R; Bubbles Photolibrary: 23; Cambridge City Council: 91ML, MR; Camera Press London: Neil Morrison 147; Camping Gaz: 97R; Colorsport: 124BL; Corbis: 127B, Neil Rabinowitz 79; Corel (NT): 13T, 16T, 40, 49, 51M, 57R, 58TL, TR, ML, 92T, 100TCL, 123M, 124TL, 125TR, 141B, 144TL, TR, BR, 145TL, TR, 166; Dan Smith: 31; Dave Rogers: 34T; Diamar (NT): 71B, Digital Vision (NT): 6T, 16M, 46, 51T, 56M, 90TR, B, 100TCR, 102T, B, 103T, 130R, 144BL; Ecoscene: Robert Gill/Papilio 124TR, Eva Meissler 168T; Empics: John Marsh 124BR, Tony Marshall 146M; F C Millington: 55T; Ford UK: 144M; Frank Lane Picture Agency: 44M, R Bender 8BR; Gene Cox: 37, 42B; Geoscience Features Picture Library: 100M, 169TL, TR, M, B; Getty Images Photodisc: 5B Ryan McKay; Getty Images Stone: P Correz 17, J Causse 18f, Dan Smith 19a, Jo McBride 19e, 38B, Bob Torrez 59, Ben Edwards 66T, Oliver Strewe 121, Chad Slattery 139; Greenpeace International: 56MTL; Holt Studios International: 44T, 52B, N Cattlin 55B, 56BL; Horticulture Research Institute: 52T, M; Hulton Getty: 106T; ICI: 48, 103B; Image Library (NT): CD5 58MR; Image State: 109; Impact Photos: 19c, 26B, 27B; J Allen Cash: 24, 25, 42T, 70B, 71T, 141TL; John Birdsall Photography: 20, 35; J P Fankhauser: 149; Kings College London: 14M; Last Resort Picture Library: 18e, 71BR, 76BR; Leslie Garland Photolibrary: 141ML; Loints Funival: 87B; Martyn Chillmaid: 5T, M, 8T, 10T, 28T, B, 33T, 44B, 51B, 53T, 56MTR, 61M, 62, 63L, M, R, 64T, B, 65TL, TM, TR, B, 66BL, BR, 68, 70TL, M, 72L, R, 75L, R, 76T, 77R, 80T, B, 83, 84M, R, 85T, 86B, 90TL, 96B, 97L, 98, 99, 108, 110, 112T, 114, 115, 122, 123B, 125TL, 142R, L, 143, 148B, 156, 157; Mary Evans Picture Library: 106M, Devany 128B; Michael Powell: 56T; Muscular Dystrophy Group: 112B; Muslim Heritage: 36T; NASA: 93, 129, 130L, 138; National Motor Museum: 144TC; National Portrait Gallery: 14T, 127M; Natural Mineral Water Information Service: 92M; Natural Visions: 41; N Downer: 56BR; New Media: 168B; Nick Cobbing: 13M, B; Oxford Scientific Films: Michael Fogden 174; Parke Davis: 36B; Panos Pictures: Trevor Page 53B, Neil Cooper 82, Trygve Bolstad 92BL; Photolibrary International: 135ML; Popperfoto: Reuters 88; Porsche: 100TL; Rail Images: 69; Raleigh Bicycles: 141TR; Robert Harding Picture Library: 87T, M H Black 9, Roy Rainford 116; Roslin Institute: 11T; Rex Features: Jonathan Player 12T; Riverford Organics: 67; Science & Society Picture Library: 107L, R; Science Photolibrary: 7B, 12B, 29TL, 34M, B, 127, CNRI 6B, TH Photo Werbung 10B, Ed Young/Agstock 11B, Vittorio Luzzati 14B, A Barrington Brown 15L, Division of Computer Research and Technology/National Institute of Health 15R, Alex Bartel 22, Department of Clinical Radiology/Salisbury District Hospital 29TCL, TCR, B, Alfred Pasieka 29TR, John Mead 39, Andrew Lambert Photography 61L, Jim Amos 70L, Charles D Winters 78, David Nunuk 81T, Martin Bond 85M, 125B, J-L Charnet 126, R Ressmeyer 134L, ESA 135T, ESC 135MC, NRSC 135MR, Cordelia Molloy 135B, Martin Dohrn 141BR; Sonia Cheadle/Studio42/33 St John's Square/London/EC1M 4DS: 77L; Sporting Pictures: 159; Sternwarte Kremsmunster: 137B; Stockbyte (NT): 100TR; T Hill: 85B, 97M; Tom Hevezi: 38T; Tony Duffy: 30; Topham Picture Point: 7T; Trip: G Horner 76BL; Trumpf International: 71MBL; Picture research by johnbailey@axonimages.com

Contents

The spice of life

Learn about:
- characteristics of offspring
- variation within and between varieties

Take a look around the classroom.
You can see that everyone has a number of different features or **characteristics.**

▶ Look at 2 of your friends and write down 3 differences in their characteristics.

▶ Look at the picture of the Green family:

a Which characteristics do you think Simon has inherited from his Mum**?**

b Which characteristics has he inherited from his Dad**?**

c Which characteristics are a result of his environment**?**

d Why do people from the ***same*** families look more similar than people from ***different*** families**?**

Coming up roses

There are lots of different varieties of roses.
They vary in their colour and number of flowers and their different heights.

e What causes their different colour and number of flowers**?**

Their different heights could be due to inherited genes.
But it could also be the result of **environmental factors**.

f Name 2 of these possible environmental factors.

g Can you think of any other variations in the roses that you can not see**?**

shrub rose | hybrid tea rose | floribunda rose

Down on the farm

Gail is a farmer. She breeds cows for their milk.

Here is Gavin, her prize bull and Daisy, her best cow:

Buttercup is the first cow bred from Gavin and Daisy.

h Which features has Buttercup inherited:
i) from Gavin**?** ii) from Daisy**?**

i What features would make a good dairy cow**?**

j What are the structures present inside the nucleus of cells that control all the features of an animal**?**

Gavin | Daisy

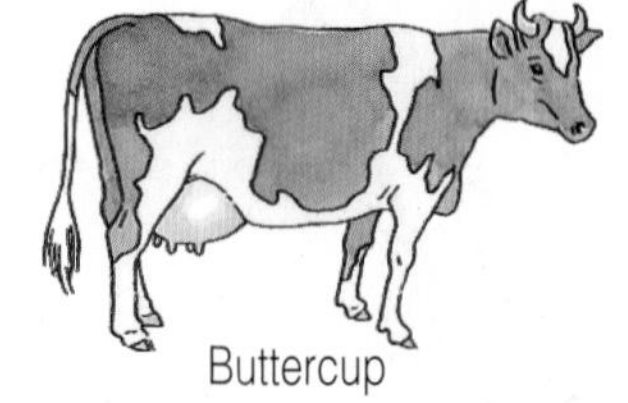

Buttercup

Variable veg!

Your teacher will give you some vegetables from the ***same*** variety of crop (they may be runner beans, French beans, potatoes or tomatoes).

Your job is to find out how they vary.
(Hint: this could be mass, length, diameter, etc.)

- Take a sample. (What would be a good sample size?)
- Make measurements of the sample and record your results in a table.
- Present your data as a frequency graph (you could use a spreadsheet).
- What conclusions can you make from your graphs?

Different varieties

This time, make the same measurements from a ***different*** variety of vegetable of the same species.

- Present your data in tables.
- Use a spreadsheet to produce frequency graphs.
- Is the variation ***within*** a variety greater or less than the variation ***between*** varieties?

  Explain how the evidence supports your conclusions.

k Many varieties differ in ways that are not observable, e.g. flavour. Can you think of any other examples?

Things to do

1 Copy and complete:
You . . . . some of your characteristics from your father and some from your . . . .
These characteristics are controlled by . . . .
Half of your genes are passed on from your father in the . . . . and half from your mother in the . . . .
Apart from your genes, your characteristics are also affected by your . . . .

2 Complete the following statements choosing the correct word:
a) Identical twins are formed from the same/different sperm and the same/different egg.
b) Non-identical twins are formed from the same/different sperms and the same/different eggs.
c) The sex of identical/non-identical twins is always the same.
d) Non-identical/identical twins look no more alike than brothers and sisters.

3 Mike and Shane are identical twins. They left home 2 years ago. Mike moved to Newcastle where he works in an office. Shane lives in Spain where he works outdoors on building sites.
a) Give 2 ways in which Shane will look different from Mike.
b) Give reasons for your answer to a).

Mike and Shane before they left home

A design for life

Learn about:
- genes and chromosomes
- how features are passed on

Have you ever heard someone say of a new baby, "Isn't she like her father?" or "Doesn't he have his mother's eyes?"

What things do you think you have inherited from your parents?

What's a chromosome?

How do we inherit things from our parents?

The **instructions** for designing a new baby are found in 2 places:

- the egg cell of the mother, and
- the sperm cell of the father.

a Which part of these cells do you think contains these instructions?

b Draw diagrams of a sperm cell and an egg cell. Label your diagrams to show how each cell is adapted for its function.

Most cells in your body contain a nucleus.
Each nucleus contains **chromosomes**.
It is the chromosomes that carry the instructions.

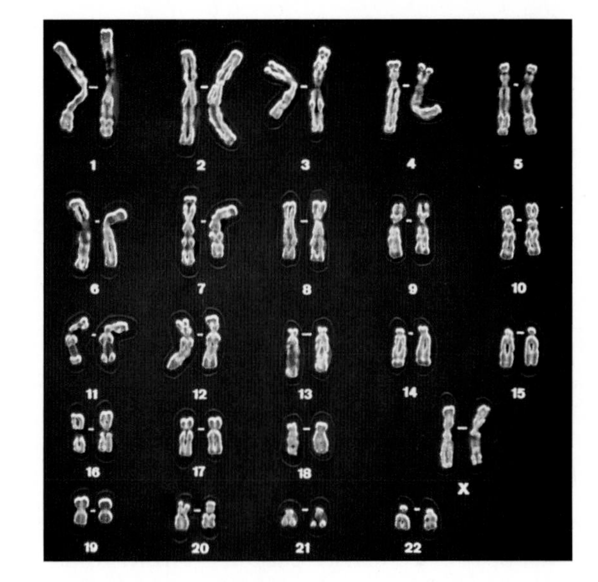

▶ Look at the photo:

c What do the chromosomes look like?

d Are they all the same size and shape?

e How many are there?

In most human cells there are 46 chromosomes.
We can put them into 23 pairs that are identical.

Halving and doubling

When a sperm cell or an egg cell is made, the chromosomes in each pair split up.

f So how many chromosomes will there be in the sperm or egg now? Do you remember what happens when a sperm fertilises an egg?

▶ Look at the diagram:

g How many chromosomes does a fertilised egg contain?

h Where have these chromosomes come from?

i What do you think would happen at fertilisation if the sperm and the egg each contained 46 chromosomes?

j Make a list of the similarities between fertilisation in plants and animals.

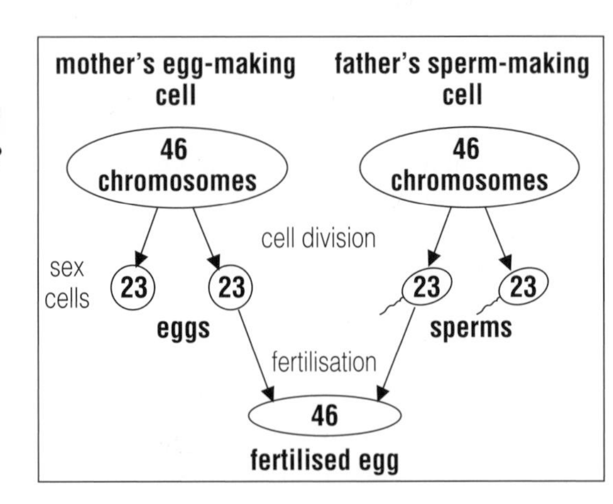

Looking at chromosomes

Scientists can make slides of human chromosomes. Your teacher will give you a sheet showing what these chromosomes look like.

- Count the number of chromosomes.
- Cut out your chromosomes and count them again to make sure you haven't lost any.
- Arrange them on a sheet of paper. Start with the largest and end with the smallest.
- Arrange your chromosomes in pairs according to size and pattern of banding.
- Stick down your chromosomes neatly in their pairs.

k Are these chromosomes from a female or a male?

l How do you know?

m What do you think the bands on each chromosome might be?

Genes

Genes contain the instructions that we inherit. Genes control certain features like eye colour.

▶ Look at the diagram:

n How many genes do you think there are for each feature?

o Where are they found?

p What do you notice about the position of genes on a pair of chromosomes?
Each band on a chromosome represents one gene.

Key to genes

- ● eye colour
- × hair colour
- ▲ tongue rolling
- ■ height
- ● nose length
- ✳ skin colour
- ■ making haemoglobin
- ∨ build of body

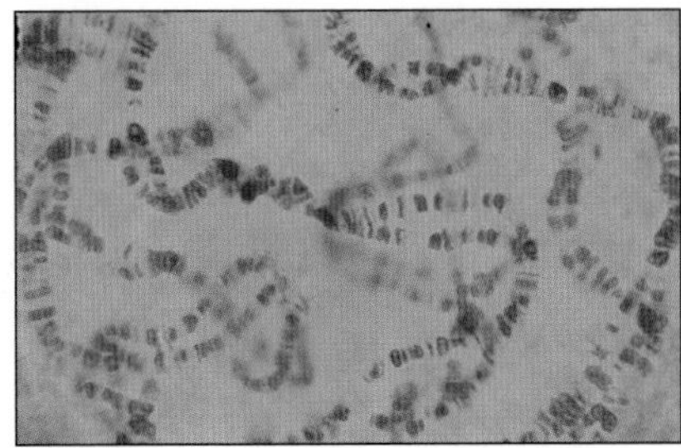

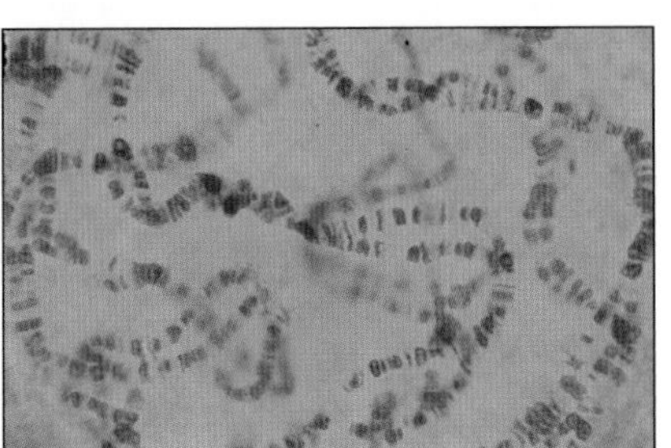

Things to do

1 Copy and complete:
Inside the nucleus of each cell there are thread-like shapes, called . . . .
These are made up of . . . .
These . . . . contain instructions to control how the . . . . works.
They also contain information which is . . . . from one generation to the next.

2 How many chromosomes are there in a human sperm or a human egg?
How many are there in other human cells?
Why do you think these numbers are different?

3 Where are genes found?
What do genes do?
Genes are made of the chemical called DNA.
Try to find out what DNA stands for.

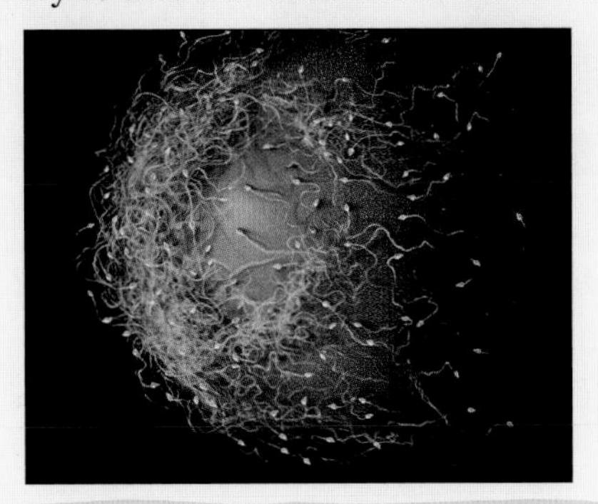

Pick and mix

Learn about:
- selective breeding
- embryo transplants

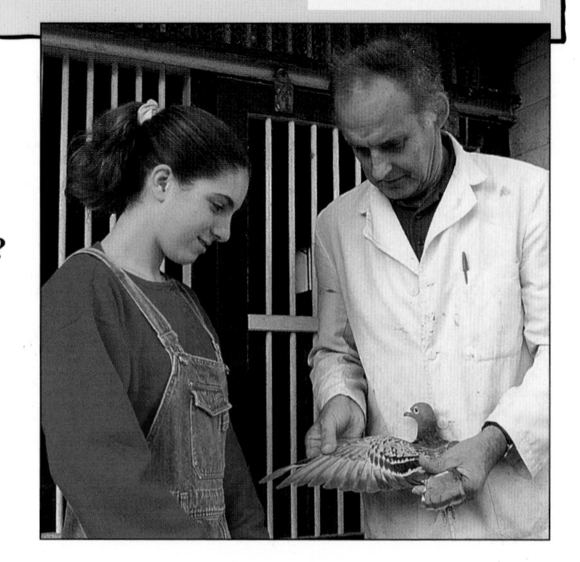

Vicki's dad keeps racing pigeons.
She helps to look after them.
They are related to wild pigeons, but have been specially bred.
They have to fly long distances but always return home.

a What features do you think have been bred into racing pigeons?

Selective breeding means that we breed in the features that we want and breed out the features that we don't want.

▶ Write down a list of animals and plants that humans have selectively bred.
Why do you think humans have selectively bred these animals and plants?

Calypso fruit

A new type of fruit has been discovered on a tropical island.
It has an amazing taste, but only when it is just ripe.
The skin is covered with hairs and it is difficult to remove.
It has a bright orange colour that often attracts birds.
The number of its seeds varies and they are very bitter.
It is said to be very good for the digestion since it contains a lot of fibre.
The calypso fruits are hard to pick because the stem has lots of thorns.

b Draw a diagram of what you think the calypso fruit looks like.
c What features would you try to breed out?
d What features would you keep?
e What would your calypso fruit look like after selective breeding?

This little piggy . . .

Our modern pig is descended from the wild pig.

▶ Look at the photos:

Bacon pig

Wild pig

f Which features do you think have been bred into the modern pig?
g Which features do you think have been bred out of it?
h Do you think the modern pig could survive in the wild?
Give your reasons.

Supercow!

Farmers can now breed better cattle.
New breeds of cattle produce more milk and more beef.

Artificial insemination means that sperm can be taken from the best bulls and put directly into the best cows.
This means that the farmer no longer has to keep bulls of his own.

i Why do you think this is an advantage to the farmer**?**

Eggs from the best cows can be removed and fertilised with bull sperm in a test-tube.
The fertilised egg is then put back into the mother cow.
These are called **embryo transplants**.

j What do you think are the advantages to breeders of:
i) artificial insemination**?**
ii) embryo transplants**?**

Carry out some research to find the answer to ***one*** of the following questions:

- Why do you think it is important to keep alive some of the old-fashioned breeds of cattle**?**
- What were the origins of domesticated farm animals**?**
- How would the 'desirable' features of sheep differ for farms in different locations**?**
- What features would you breed into poultry used for different purposes**?**

Your answer should be supported by pictures or diagrams.

Things to do

1 Make a list of 5 animals or plants that you think have been 'improved' by selective breeding. For each one, say how you think this is useful to humans.

2 A lot of cereals originally came from the Middle East.
What features do you think have been bred into varieties that are grown in Britain?

3 What features do you think have been bred into these dogs:
a) pekingese? b) sheep dog?
c) bloodhound?

4 Humans have selectively bred modern varieties of wheat from wild wheat.
These have greater yield, improved disease resistance and ripen over a shorter time.
How do you think these 3 features have helped the farmer?

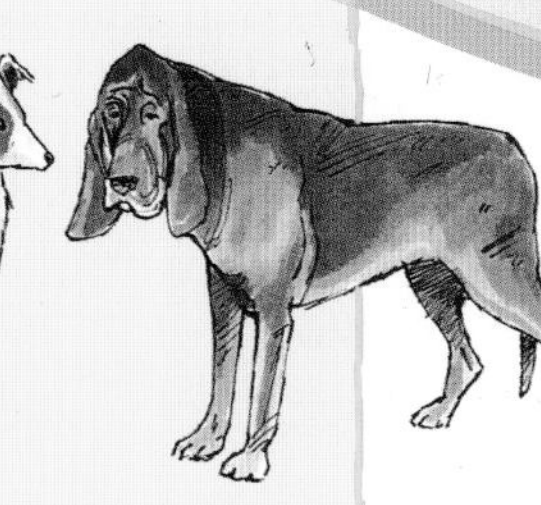

Selecting and cloning

Learn about:
- producing new variations
- selective breeding
- cloning

Over many years, plant breeders have selected healthy plants to breed the characteristics that they want.

▶ Look at the photo of the different varieties of tomato:

a What characteristics can you see that make them different? Are there any other differences that you cannot see?

b Why are each of the following characteristics important:
- resistance to the cold,
- sweetness, texture and colour of fruit,
- a long shelf life.

A common ancestor

▶ Look at the diagram below.
It shows a variety of vegetables that have been selectively bred from a single species.

c For each vegetable, say which parts have been selectively bred.
For example, the cabbage has been selectively grown for its large bud.

Peas please

Your task is to investigate the effects of selective breeding on plants.

To do this, compare the characteristics of ***fresh peas*** and ***frozen peas.***

- Think about a question or questions to investigate, e.g. do frozen peas weigh less than fresh peas?
  You might think about size, colour, taste or cooking time.
- What sort of sample size will you choose?
- Choose a method that fits your question(s).

You could pool your results with other groups before drawing your conclusions.

Then, evaluate the strength of your evidence.

Cloning

Cloning is a type of reproduction that does not involve sex cells.
We call it **asexual reproduction**.
Cloning produces individuals that are **genetically identical**.
Asexual reproduction has been used for many years to produce new plants. Here are some examples:

Cuttings can be put into rooting powder. This contains a hormone which encourages roots to grow.

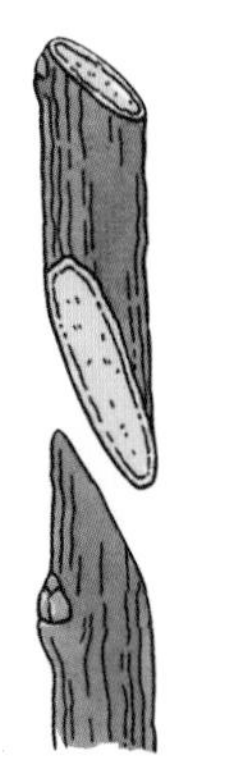

Grafting involves tying the stems of 2 similar plants together. The wound soon heals to give a single plant. Grafting is often used to produce ornamental bushes and fruit trees.

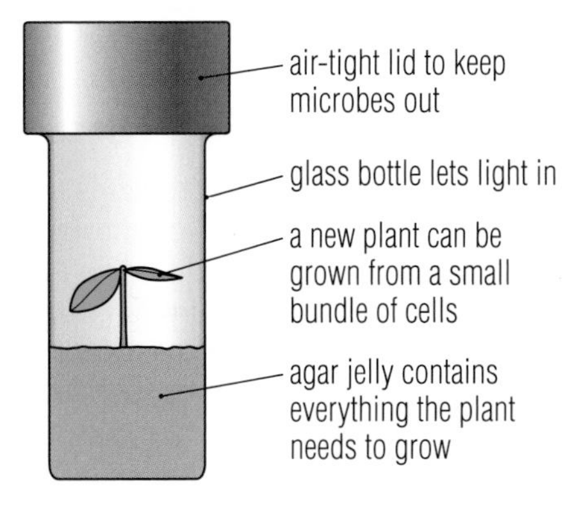

Tissue culture involves taking small pieces of tissue from a parent plant and growing them in sterile agar jelly.

Because the offspring are genetically identical to the parent plant, desirable characteristics can be passed on.

d Can you think of any examples of these desirable characteristics**?**

Research activity

Think of 5 questions to ask a scientist about cloning.
Explain why you think that your questions are important.
Use the internet to research your answers.
Discuss your answers with your friends and make a brief summary.

Your teacher can give you a Help Sheet to show you how Dolly the sheep was cloned.

What are your views on animal cloning**?**

e Can you think of any examples when animal cloning should be used**?**
(Hint: think about endangered species or raising successful farm animals in a developing country.)

Things to do

1 Copy and complete:
Genetically . . . . organisms are called . . . .
Cloning is a type of . . . . reproduction that does not involve . . . . cells. Examples of the use of cloning in horticulture include taking . . . . , tissue . . . . and . . . . Tissue culture involves growing . . . . from the . . . . plant in . . . . conditions.

2 Use a gardening or botany book or the internet to find out how asexual reproduction takes place in
a) strawberry runners b) iris rhizomes
c) potato tubers d) daffodil bulbs.

3 Here are some features about plant tissue culture. In each case say what are their advantages to the grower:
a) quick to carry out
b) does not take up too much space
c) can be carried out in any season
d) produces lots of plants from one parent
e) plants are genetically identical.

Biology at Work

Learn about:
- The Human Genome Project
- In vitro fertilisation
- Genetically modified foods

The Human Genome Project

The Human Genome Project is a six-billion dollar venture involving over 1000 scientists from 50 countries around the world. Its aim is to trace every single human gene and find its particular position on a chromosome.

a What is a gene and what does it do**?**

Genes are found on chromosomes.

b What are chromosomes**?** (Hint: see page 6)

c Where are chromosomes found**?**

Dr John Sulston, director of the Sanger Centre where the British genome work was based

In June 2000, the human genome was completed:
3 billion 'chemical letters' that spell out all the human genes.
Scientists will be able to use the information to understand how people are affected by certain diseases and target early treatment. Cancer scientists have begun to catalogue the DNA changes in cancer cells in the hope of developing totally new treatments.

d How do cancers form in the body**?**

e How do cancer cells spread through the body**?**

f Give 2 ways in which cancers are currently treated**?**

The Human Genome Project will help create new drugs to treat heart disease, immune disorders, muscular dystrophy, birth defects and degenerative nervous diseases.

g Find out more about some of these diseases.

h Use the internet to find out more about the Human Genome Project.

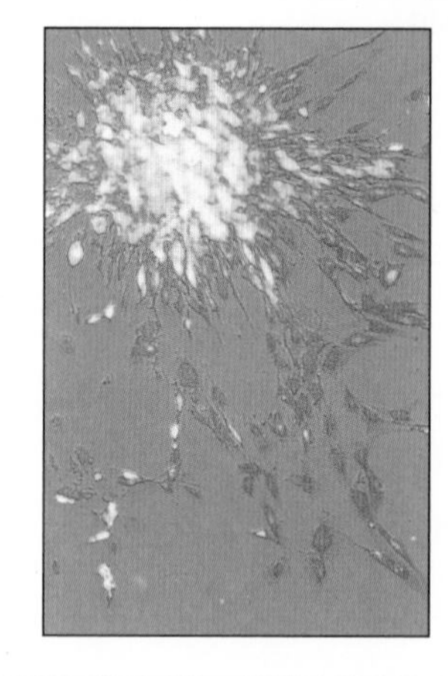

Cancer cells breaking away from a tumour

In vitro fertilisation (IVF)

Explain how each of the following could stop a couple having children:

i The man cannot make enough sperms.

j The woman's egg tubes can become blocked.

In vitro fertilisation can often help these couples. ('In vitro' means 'in glass'.)

k What is meant by fertilisation**?**

l Why do you think this technique is called IVF**?**

First the woman is injected with hormones to make her produce eggs.

m How do you think the eggs are removed from her body**?**

The eggs are then kept in a solution containing food and oxygen at the correct temperature.

n Why do you think this is**?**

Look at the diagram:

o Write a paragraph to explain exactly how IVF is carried out.

Sometimes more embryos form than can be used.
Many people think that it is wrong to destroy these extra embryos.

p What do you think**?** Write about your thoughts.

Sometimes these extra embryos are frozen. They can then be used later if the first embryos do not grow.

q Do you think that it is right to do this**?** Write about your thoughts.

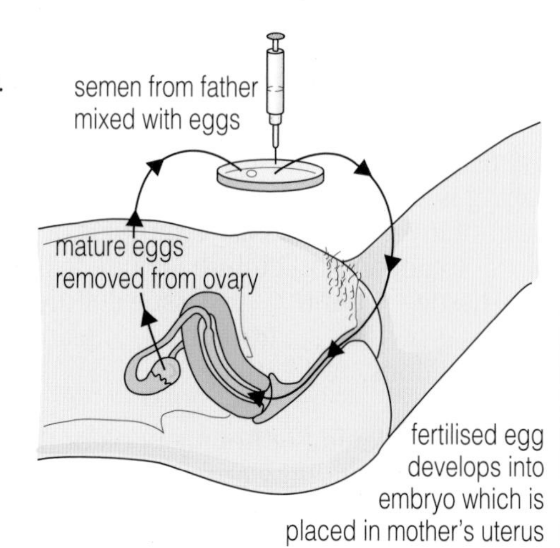

The GM food debate

Genetically modified (GM) food in the form of soya products, tomato purée or vegetarian cheese has probably been eaten by most people already.

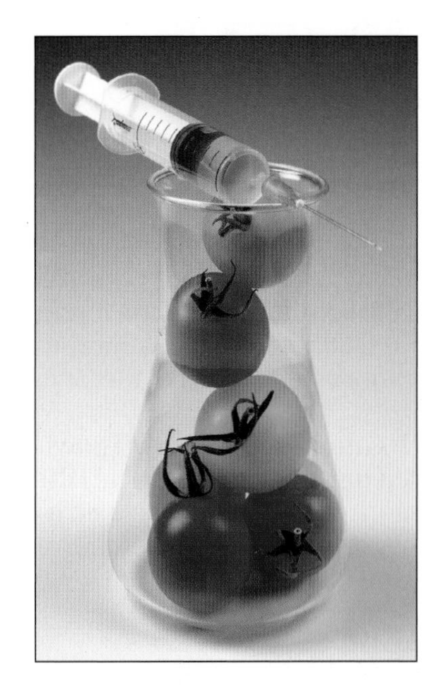

What are GM foods?

Genetic engineering has allowed sections of DNA to be removed from some plants and transferred into other plant cells.

r What chemicals are used to remove and transfer this DNA?

GM foods – the benefits

- **Solving global hunger** – crops can be developed with the genes that give them tolerance of drought, frost and salty soil.

s How does this help solve food shortage?

- **Environmentally friendly** – genetic modification can give crops resistance to insect pests, weeds and diseases.

t How can this help the environment?

Crops can be developed that are able to take up nitrogen better.

u How can this help the environment?

- **Consumer benefits** – GM food can be produced with better flavour, better keeping qualities and which need fewer additives.

v How do these qualities help suppliers and consumers?

GM foods – the concerns

Opposition to the increased use of GM foods has come from the following areas:

- **Environmental safety** – there are worries that GM crops may become successful weeds and transfer their genes to related plants.

w How might a plant that develops insect-resistant genes become a weed?

- **Food safety** – there are concerns that the proteins that GM food genes can produce could be transferred to microbes in the human intestine.

x How might this prove to be a threat to human health?

- **Changes in farming structure** – there has been a trend, over the past decades, towards larger more intensive farms.

y Why might the production of GM foods favour wealthy farmers?

- **Biodiversity** – the control of plant breeding could result in fewer varieties of plant species being available to farmers.

z Why could the reduced use of old varieties and their wild relatives put more plants in danger of attack by pests and diseases?

- **Animal health** – there is increasing resistance to any developments in the use of farm animals that could affect their welfare.

A **Either** research further aspects of the benefits and concerns of GM foods on the internet **or** set up a debate with groups arguing these benefits and concerns about this new biotechnology.

Greenpeace activists destroying a GM trial crop

Variation

Learn about:
- Discovering the structure of DNA
- How new ideas arise in science

Ideas about DNA

Chromosomes are always found inside the nucleus. When a cell divides, they split lengthways and provide an identical copy of each chromosome for each of the 2 new cells.

Each chromosome is now known to be made up of an extremely long coiled thread of the chemical **DNA.**
Sections of the DNA making up a chromosome are called **genes.**
Genes provide the code for the formation of certain characteristics, e.g. eye colour.

In the 1950s, a great deal of work was being carried out by scientists to discover the nature of DNA.

Erwin Chargaff, the American biochemist, analysed samples of DNA from different organisms and made conclusions about its chemical make-up.

In 1952, **Frank Hershey** and **Martha Chase** proved that it was the DNA in viruses that controlled the characteristics of each virus.

Maurice Wilkins

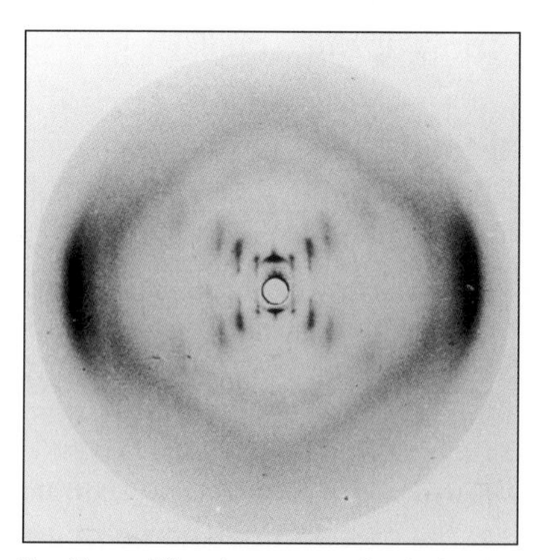

The X-ray diffraction pattern that helped provide clues about the structure of DNA

X-ray diffraction techniques

In 1953, **Rosalind Franklin** and **Maurice Wilkins** at King's College, London, used X-ray diffraction to work out more about how the atoms inside the DNA molecule were arranged.
This technique involved firing a beam of X-rays into crystals of DNA. The X-rays hit the atoms and were scattered, making up a pattern on a photographic plate.

The X-ray diffraction photos taken by Rosalind Franklin helped to build up a picture of what the DNA molecule was like.
Wilkins showed one of Franklin's pictures to James Watson, an American biochemist, working in Britain.
He immediately realised that it held the answer to the structure of DNA.

In 1962, Watson along with his co-worker Francis Crick and Maurice Wilkins, were awarded the Nobel Prize for their work on DNA structure. However, Rosalind Franklin had died of cancer 4 years earlier.
Whether she would have shared the Nobel Prize if she had lived longer, we cannot say.

Rosalind Franklin

The double helix

The molecular structure of DNA was finally worked out by **James Watson** and **Francis Crick**, working in Cambridge in 1953.
They used all the latest information including Chargaff's results and the X-ray diffraction pictures of Franklin and Wilkins.
This meant piecing together cut-out models of the molecules involved and fitting them together.
After months of discussion and painstaking manipulation of the model, they were able to build a 3-D model of the structure of DNA.
It turned out to be a beautiful 'double helix' shape.
Each helix could separate from the other and make an exact copy of itself. This meant that chromosomes and genes could also make exact copies of themselves when cells divided.

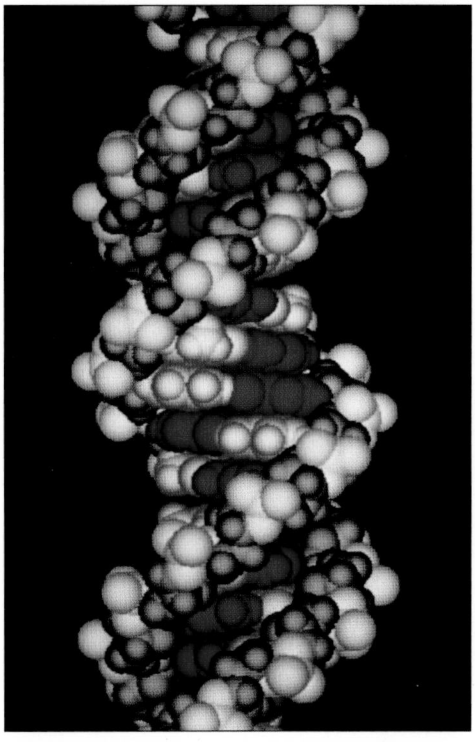

Computer generated model of part of a DNA molecule

Watson and Crick with their DNA model

Questions

1 Explain what is meant by each of the following words:
a) chromosome b) DNA c) gene

2 a) What controls inherited features?
b) Give 3 examples of features that are controlled by genes.
c) What features are controlled by the way we lead our lives?
d) Give 3 examples of features that are affected by our environment.

3 Why do you think so many scientists were working to discover the nature of DNA in the early 1950s?

4 Explain the X-ray diffraction techniques used by Franklin and Wilkins.

5 What is meant by the phrase 'double helix'? Why is it sometimes compared with a spiral staircase?

6 Why was it important that any model of DNA should be able to make an ***exact*** copy of itself? (Hint: think about what happens to chromosomes when a cell divides.)

7 Try to find out more about the work of:
a) Hershey and Chase, or
b) Franklin and Wilkins, or
c) Watson and Crick.
You could use books, ROMs or the internet.

Questions

1 Say if you think the following statements are true or false:
a) Chromosomes in sex cells occur in pairs.
b) At fertilisation, half the chromosomes come from the father and half from the mother.
c) Chromosomes are ***only*** found in the nucleus of a cell.
d) Blood groups and eye colour are ***not*** inherited characteristics.
e) Genes are sections of DNA found along chromosomes.

husky

2 Look at the list of working dogs below and say what characteristics have been bred into each one to make it good at its job.
a) Husky
b) Bloodhound
c) Collie sheepdog
d) Labrador guide dog

3 Breeding racehorses is a million-pound industry.
What sort of features do you think would be bred into a champion racehorse?

4 Cloning involves taking some pieces of tissue from a parent plant and growing them in sterile conditions in nutrient agar.
a) Why will the cloned plants be 'genetically identical' to the parent?
b) Why are the clones grown in sterile conditions?
c) Explain how plants grown in this way can:
i) be produced quickly ii) be free from disease iii) have the 'good' qualities of the parent plant, e.g. disease-resistance.
d) Find out and write about any examples of animal cloning.

5 If animals of 2 different species mate they produce a **hybrid**.
The tigon resulted when a tiger and a lion mated in captivity:
a) What do you think were the parents of a leopon and a zeedonk?
b) Invent your own animal from 2 different species and draw it.
c) Hybrids are usually sterile (they cannot produce offspring). Why do you think that the survival of some species would be threatened if they bred with other species in the wild?

tigon

6 Look at the table of 2 breeds of sheep:

Feature	*Welsh Mountain breed*	*Border Leicester breed*
adult mass (kg)	30–32	73–77
number of lambs born per female (ewe)	1	2
temperament	wild	docile
fleece per ewe (kg)	1	3
growth rate	slow	fast

a) What features do you think have been bred into the Border Leicester?
b) How do these make it a successful sheep?
c) Hill farmers often cross Welsh Mountain ewes with Border Leicester rams. Why do you think this is?

Fit and healthy

We all want to have happy, healthy and active lives. But to do so, we need to take good care of our bodies. We should eat the right foods and take regular exercise. We should understand the dangers to our health of alcohol, drugs and solvents.

In this unit:

Fit for life?

Learn about:
- different types of fitness

Are you fit?

▶ Write down some of your ideas about what it means to be fit.

Look at the people in the photos:
Each person must be fit to do their sport well.

S-factors

Four things make up fitness:

  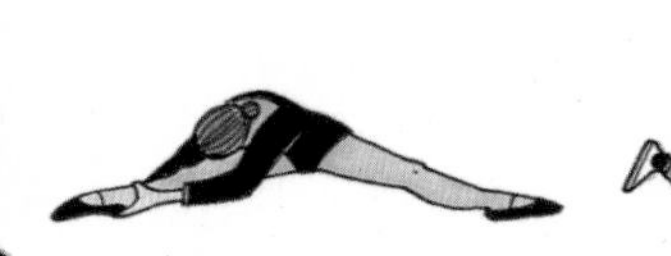 

- **Strength** is the amount of force that your muscles can exert.

a Which exercises do you think could make you stronger?

b Which things that you do every day need strength?

c Which sports in the photographs need strong muscles?

- **Stamina** keeps you going during exercise.
  To develop your stamina, you need a strong heart and lungs.

d Which sports in the photographs need a lot of stamina?

e What sort of exercises do you think would improve your stamina?

f How can people's life-styles reduce their stamina?

- **Suppleness** lets you move freely and easily.
  If you are supple, you can bend, stretch and twist your body easily.

You often see people 'warming up' before they do sport.

g What exercises do they do?

h What might happen if they did not do these 'warm up' exercises?

i Which sports in the photographs need you to be supple?

- **Speed** is having quick reactions or how fast you travel over a distance.

j Which sports in the photographs need speed over a distance?

k Which sports in the photographs need quick reaction times?

l How can you improve both types of speed?

We're getting fitter!

More people are getting exercise from sport or fitness activities. When you do different activities you need different amounts of your S-factors.

▶ Look at the table.

Exercise	*Strength*	*Stamina*	*Suppleness*	*Speed*
Badminton	**	**	***	**
Climbing stairs	**	***	*	*
Cycling (hard)	***	****	**	*
Dancing (disco)	*	***	***	*
Football	***	***	***	***
Golf	*	*	**	*
Gymnastics	***	**	****	**
Hill walking	**	***	*	*
Jogging	**	****	**	*
Swimming	****	****	****	**
Tennis	**	***	***	***
Weight-training	****	*	*	*

*no real effect **good effect ***very good effect ****excellent effect

m Which exercises do you think are best for i) strength**?** ii) stamina**?** iii) suppleness**?** iv) speed**?**

n Which exercise do you think is i) best and ii) worst for your all round fitness**?**

o Karen is 14. She isn't sporty. What exercises can you suggest to keep her fit**?**

Fitness programme

Think up a fitness programme that a year 9 pupil could do in 10 minutes a day. Make a leaflet of your programme. It should:

- include all four S-factors (strength, stamina, suppleness, speed).
- not involve any special equipment like weights.
- not be too difficult.
- not need a large space to do it in.

You should not try the programme out unless it has been checked by your teacher. And remember, if you feel any strain during exercise, stop and rest.

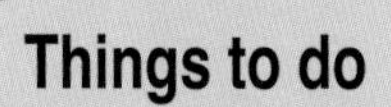

1 Copy and complete:
Strength is the amount of . . . . that your . . . . can exert . . . . . keeps you going when you exercise hard. If you can bend, twist and stretch your body easily, then you are . . . . Speed can mean being fast over a . . . . or having quick . . . . You need all 4 of these . . . . to be fit.

2 Think up a fitness programme for someone who does a) netball b) rowing c) sprinting.

3 Use the table above to suggest suitable exercises for these people:
a) a 45-year-old man who has recovered from a heart attack.
b) a 27-year-old mother who had her first baby 10 weeks ago.
c) a 50-year-old woman who has never played sport.

4 Choose one sport that you enjoy and design a poster to encourage people to take part in it.

Are you healthy?

Learn about:
- energy input and output
- respiration

What do you think we mean when we say that someone is **healthy?**

▶ Write down some of your ideas about health.

A scientist once described health as:
'the state of complete physical, mental and social well-being'.

a Under the headings 'physical', 'mental' and 'social', make a table of the different things that can affect a person's health.

A healthy diet

In the last lesson we saw how important it is to exercise to be healthy. But you also need the right sort of food to enable you to exercise.

b Write down the foods that make up a **balanced diet**. Remember that you need ***enough*** of each of these foods.

c Which of these foods are ***energy*** foods and which are ***growth*** foods?

You need to balance your **energy input** with your **energy output**.

d What happens if your energy input is greater than your energy output?

e What happens if your energy output is greater than your energy input?

Remember we release energy from food in the process of **respiration**:

GLUCOSE + OXYGEN ⟶ CARBON DIOXIDE + WATER + ENERGY

f Which body systems are involved in providing our cells with glucose and oxygen, and getting rid of the waste products of respiration?

A healthy lifestyle

Sometimes people do things that have a bad effect upon their health.

g For each of the following habits, explain the bad effects that they can have upon a person's health:

smoking cigarettes ***drinking alcohol*** ***eating fatty foods***
solvent abuse ***taking harmful drugs*** ***not exercising***

We will be looking at some of these harmful habits in the next few lessons.

▶ Discuss with a friend and write down your ideas about how some of the following can help to give you a healthy outlook on life:

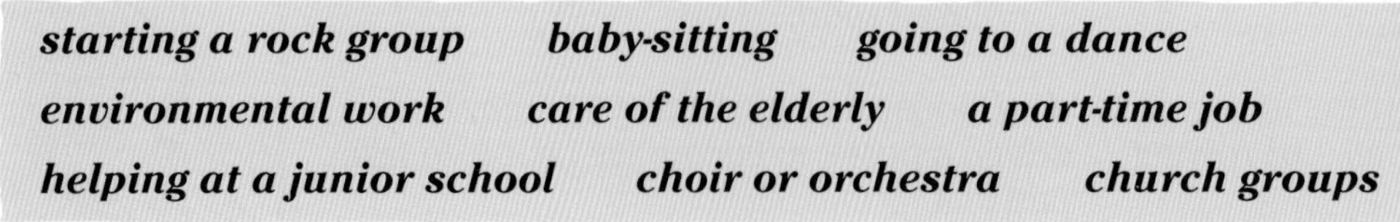
starting a rock group ***baby-sitting*** ***going to a dance***
environmental work ***care of the elderly*** ***a part-time job***
helping at a junior school ***choir or orchestra*** ***church groups***

Lots of clots

For healthy growth you need a balanced diet.
You especially need protein and vitamins for growth.

When you were a baby you got your protein from milk.
The enzyme **rennin** is made in the stomachs of young animals.
It makes milk solid, so that it stays in the stomach longer.
It can then be digested.

Plan an investigation into how quickly rennin clots milk.
Think about the factors that might affect how quickly rennin works.
Choose one factor and investigate its effect.

- How will you make it a ***fair test*?**
- How will you decide when the milk has clotted**?**
- How will you record your results**?**

Show your plan to your teacher before you try it out.

Health and disease

As we have seen, you don't have to have a disease to be unhealthy.
But lack of health can be caused by disease.

Some diseases are caused by **germs**, such as bacteria, viruses and fungi.

h Find out the names of 3 diseases that are caused by bacteria.
i Find out the names of 3 diseases that are caused by viruses.
j Find out the names of 2 diseases that are caused by fungi.

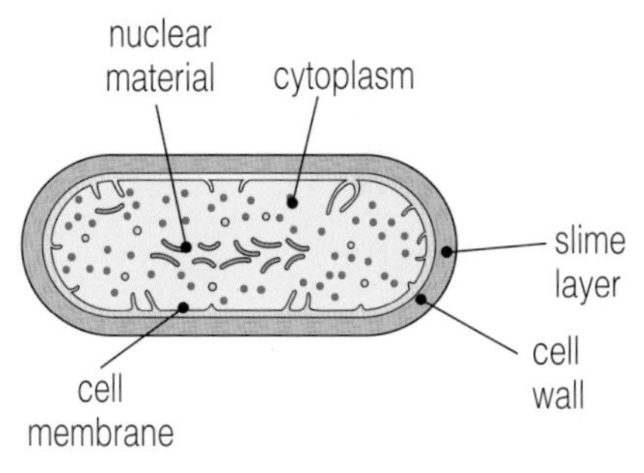

The basic structure of a bacterium

Some diseases are ***not*** caused by germs (**microbes**).
Diabetes, arthritis, cancers, cystic fibrosis and motor neuron disease are ***not*** caused by microbes.

Find out about these diseases from books, ROMs and the internet.

Things to do

1 Here is a table which shows Liam's height since he was born:

Age (years)	birth	2	3	4	5	6	7	8	9	10	11	12	13	14	15	16	17	18
Height (cm)	50	85	95	103	110	115	120	125	130	135	140	145	150	160	170	173	174	175

a) Plot a line-graph to show how Liam's height has changed.
b) Explain the shape of the graph.
c) During which years was Liam growing fastest?

2 Copy and complete:
Health can be described as 'the state of complete physical, ... and . . . . well-being.' A . . . . diet combines the correct types of foods in the right . . . . This should include . . . . and fats for energy and proteins which are needed for . . . . and the repair of cells. Too much intake of fatty foods can lead to problems in the . . . . system. Certain nutrients such as . . . . and . . . . are needed in small amounts in the diet, but if they are absent, they can result in . . . . diseases.

3 a) Make a list of habits that can have a bad effect upon the health of a person.
b) What particular advice would you give to i) teenagers ii) 30 year olds and iii) pensioners, about how to maintain a healthy lifestyle.

4 There has been a lot of coverage in the press about athletes and performance-enhancing drugs. Use books, ROMs and the internet to write about i) whether this is right or wrong ii) how these drugs improve performance and iii) the bad effects that these drugs can have on an athlete's body.

Hormones also control your growth

Dying for a smoke?

Learn about:
- harmful substances in smoke
- effects of smoking

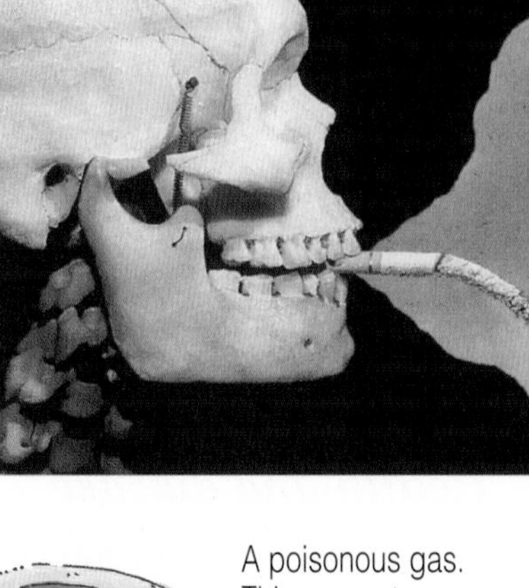

Do you know anyone who smokes? There are fewer smokers around these days. Lots of people used to smoke but nowadays they are finding it less attractive.

▶ Write down your ideas about why this is.

Did you know that cigarette smoke is made up of lots of chemicals and many of these are poisonous? If you smoke, these chemicals go into your body through your mouth and along your air passages.

Remind yourself how your respiratory system works. (See 8B2 and 8B3 in Book 8.)

▶ Explain how we breathe in and out.

Fancy this lot?

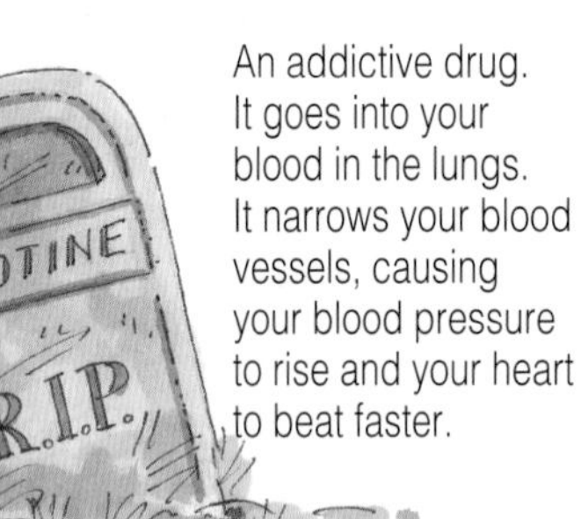

An addictive drug. It goes into your blood in the lungs. It narrows your blood vessels, causing your blood pressure to rise and your heart to beat faster.

A brown, sticky substance that collects in your lungs if you breathe in tobacco smoke. It is known to contain substances that cause cancer.

CARBON MONOXIDE
R.I.P.

A poisonous gas. This prevents your blood from carrying as much oxygen as it should and so you get out of breath easily.

The smoking machine

First set up the apparatus without the cigarette.

Turn on the suction pump.

After 5 minutes, record:

- the temperature
- the colour of the glass wool
- the colour of the lime water.

Now repeat the experiment with a cigarette.

Record your observations.

What does this experiment tell you about the difference between breathing in fresh air and cigarette smoke?

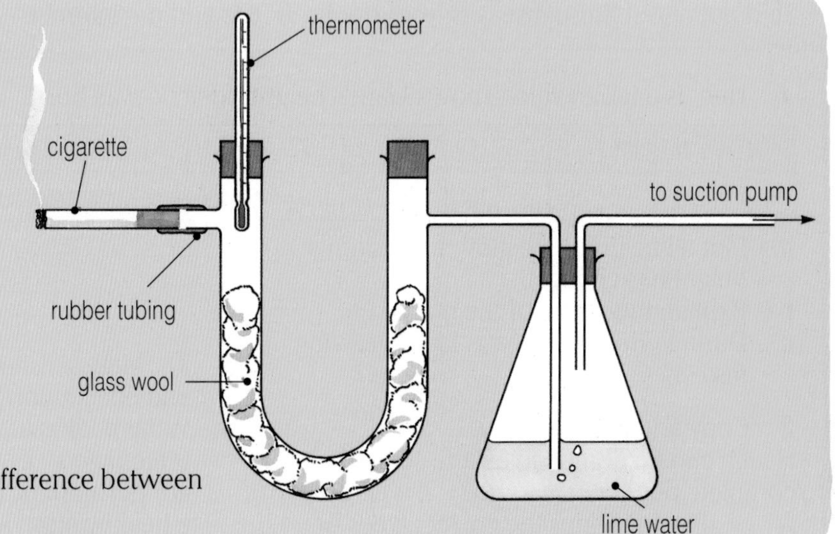

Smoking changes people

hair and clothes smell – that's the smoke

tongue turns yellow – you can't taste food properly

teeth, fingers and nails turn yellow – that's the nicotine

smoker's cough – to get rid of the mucus (this is because heat affects the way hairs on the ciliated epithelial cells work)

carbon particles in the smoke irritate the lining of the alveoli

A mug's game

▶ Read what some people say about smoking.
Design a leaflet for primary school children to explain why they should not start smoking.

Ninety per cent of lung cancer occurs in smokers.

Smokers are 2 to 3 times more likely to die of a heart attack.

The money spent on cigarettes can't be spent on food, clothes etc.

Smoking increases the risk of serious diseases like bronchitis.

Smoking makes you breathless and less good at sport.

You can't keep the smoke to yourself. Everyone around you has to breathe it in.

Illnesses caused by smoking have to be treated. If people did not smoke, this would cost the country less money.

The substances in smoke inhaled by pregnant women affect the development of the fetus.

So why start?

Amy is 13. She smokes about 5 cigarettes a day.

▶ Read what Amy has to say about smoking:

"I had my first cigarette when I was 10.
My friend Sharon, she's 2 years older than me, offered me one.
I didn't like it much at first, but it felt exciting somehow.
My Mum would have killed me if she'd found out.
She's always trying to get my Dad to give up but he can't.
I suppose I spend about £4 or £5 a week on cigarettes.
I'll give it up when I'm older because it affects your health.
I'd definitely give up if I ever got pregnant."

In your groups, discuss the reasons why you think people start to smoke.

Things to do

1 Copy and complete:
Cigarette smoke contains poisonous . . . . One is a drug called . . . . This gets into your blood in the . . . . It causes your blood . . . . to rise and your heart to beat . . . . Tar contains chemicals that cause . . . . A poisonous gas called . . . . stops your blood from carrying as much . . . . as it should.

2 You are asked to talk to some junior school children about the dangers of smoking.
Plan out what you are going to tell them.

3 Write down what you think about the following:
a) Smoking should be banned in shops, offices and on public transport.
b) Once you start smoking it's hard to stop.
c) Smoking costs us all a lot of money.

4 The graph shows how the risk of getting lung cancer changes after giving up smoking. Explain why you think smokers should give up the habit.

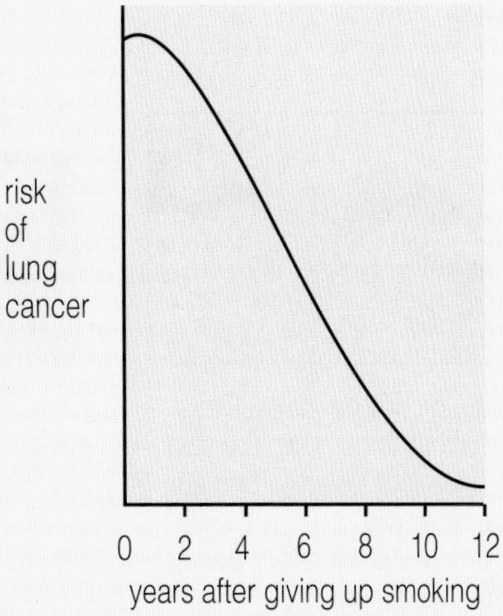

risk of lung cancer

0 2 4 6 8 10 12

years after giving up smoking

A drink problem

Learn about:
- the effects of alcohol

Alcohol is a **drug**.
In Britain drinking alcohol is **socially acceptable** but people can become **addicted** to it.

a Write down what you think the highlighted words mean.

Most drugs affect the brain and nervous system.
Alcohol is a **depressant** drug.

b What effect do you think alcohol has on the way your body works?

Alcohol is made by **fermentation**.

c Can you remember what happens during this process?

d Alcohol is sipped and swallowed. But where do you think it goes after that?
Your teacher may give you a Help Sheet which shows how alcohol can affect your body.

Units of alcohol

All these drinks contain **1 unit** of alcohol:

After drinking 1 unit of alcohol, the amount of alcohol in the blood increases by 16 mg in 100 cm^3.

▶ Match **e** to **j** with the correct units of alcohol in the pictures.

e The legal limit for driving.

f No obvious effects, but your reactions are slower.

g Speech is slurred, seeing double, feeling emotional, may be tearful or looking for a fight.

h Talkative, your judgement is not so reliable.

i Possible loss of consciousness.

j Feeling more cheerful.

1 unit

12 units

2 units

16 units

3 units

5 units

Dave goes downhill

When Dave started work for a local firm he soon got to know the other lads. He had the odd drink with them at lunchtime even though it made him feel a bit sleepy.

With the money he made, Dave could afford to go down the pub in the evenings as well. Some nights he would stay out so late that he found it difficult to get up for work next morning.

One afternoon Dave made a mistake that could have caused a serious accident at work. As he had been warned about being late many times, this was the last straw – he was sacked!

Out of work and with nothing to do, Dave now needed a drink even more. He became bad-tempered and started to borrow money from friends for a drink.

k What do you think was the cause of: i) Dave's mistake at work? ii) Dave being late for work?

▶ Look at the graph:

l How many units of alcohol are removed from the blood every hour?

m Dave goes home at 11 p.m. after drinking 10 pints of beer. How long does it take for all the alcohol to get out of his blood?

n At what time would he be below the legal limit for driving?

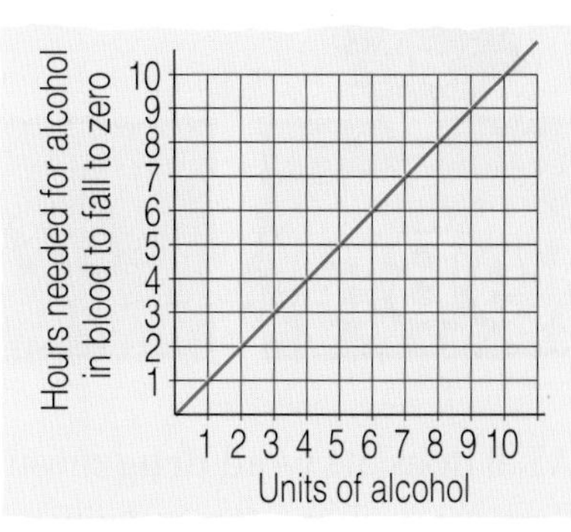

What people say about drinking

In your groups discuss:

- Why you think each person drinks.
- Which people you think are at low risk from alcohol.
- Which people you think could become problem drinkers.

Either Choose to be one of the characters and act out a discussion about the dangers of heavy drinking.

or Choose a character and write a story to show how heavy drinking can cause difficulties in the family.

Things to do

1 Copy and complete:
Alcohol is a . . . . because it affects your . . . . system by . . . . down your reactions. For this reason, people should not drink if they are to . . . . machines or . . . . a car. A pint of beer contains . . . . units of alcohol. The legal limit for driving is . . . . People who drink too much can become . . . . to alcohol.

2 When Louise was 13 she took a friend home at lunchtime. They helped themselves to drinks from her parents' drinks cabinet. Write an ending to the story.

3 What do you think should be done to get rid of the problems caused by alcohol? How successful do you think these would be:

a) pubs opening for longer or shorter times?
b) raising or lowering the age at which it is legal to buy alcohol?
c) making alcoholic drinks more expensive?

Linda and Carl

Learn about:
- the effects of drugs

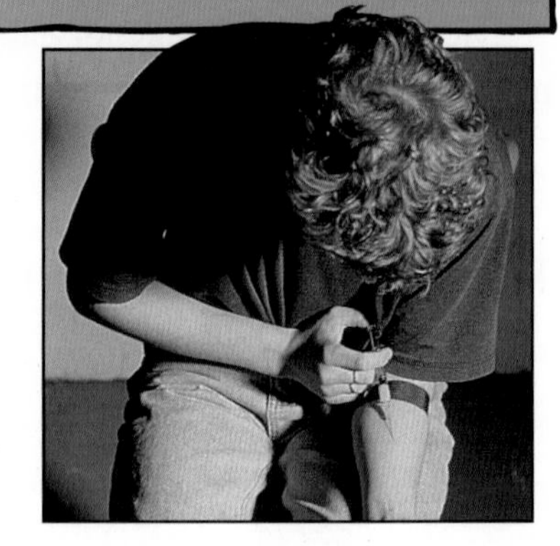

What does the word **drug** mean to you?

Here are some pictures of different drugs.

▶ Use them to help you write about what a drug is.

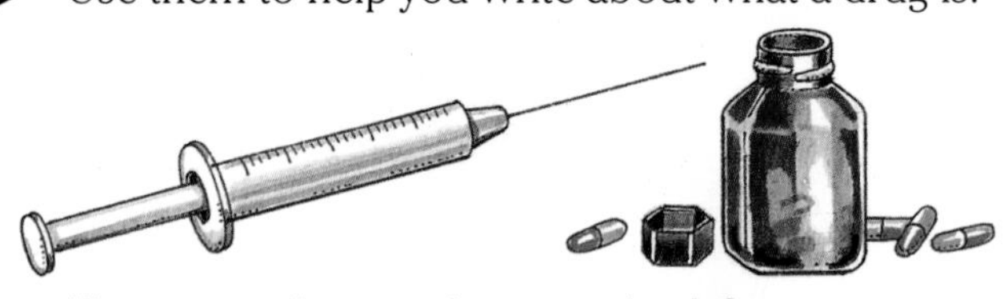

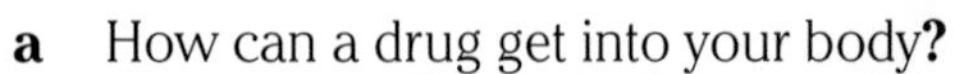

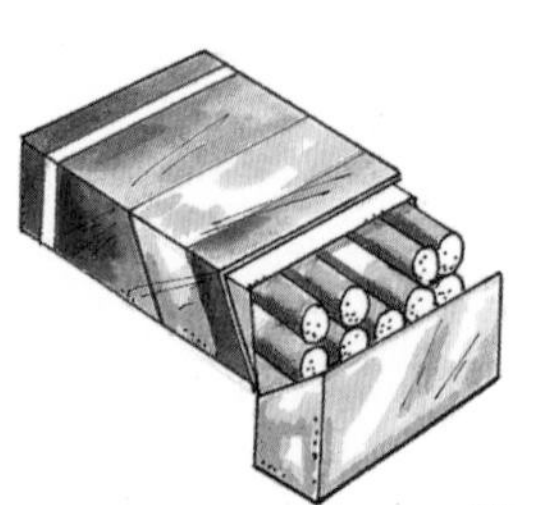

a How can a drug get into your body?

b Why can it affect ***all*** of your body?

c How do you think a drug can save lives?

d How many drugs do you know of? Make a list of them.

e Did you include any of these in your list: aspirin, coffee, alcohol, cigarettes, insulin, aerosols? Why do you think that these can be drugs?

f Classify the drugs from **d** and **e** into these groups: prescription drugs; over-the-counter drugs; legal recreational drugs; illegal drugs. Which drugs were difficult to classify? Why?

Disco girl killed by one tablet of new drug

An 18-year-old girl has died after taking a single tablet of a new designer drug. A court heard yesterday how Linda took the pill to give her energy at an all-night disco. She collapsed on the dance floor and was rushed to hospital screaming in pain. She suffered 2 heart attacks and died from lack of oxygen to the brain 2 days later. Her parents were at her bedside.

The judge branded the drug barons who supply drugs like ecstasy "scum and filth". He imposed 6-month sentences on 3 young people who admitted pushing the drug.

In your groups discuss:
- Why Linda took the drug in the first place.
- How she got hold of the drug.
- What action should be taken, and by who?

In your groups: Talk about situations where teenagers may be offered drugs. What reasons might influence them to accept or refuse?

and Write a script for a short scene showing the dangers of taking drugs. Choose a role and try to speak and act as if you are the person. Your teacher might choose your group to act your scene to the class.

or Write one or two paragraphs saying why you think teenagers start to take drugs. Say what your views on drug-taking are.

Solvent abuse

What do you think is meant by **solvent abuse?**

▶ Use your ideas and some of these facts to write a few lines about it.

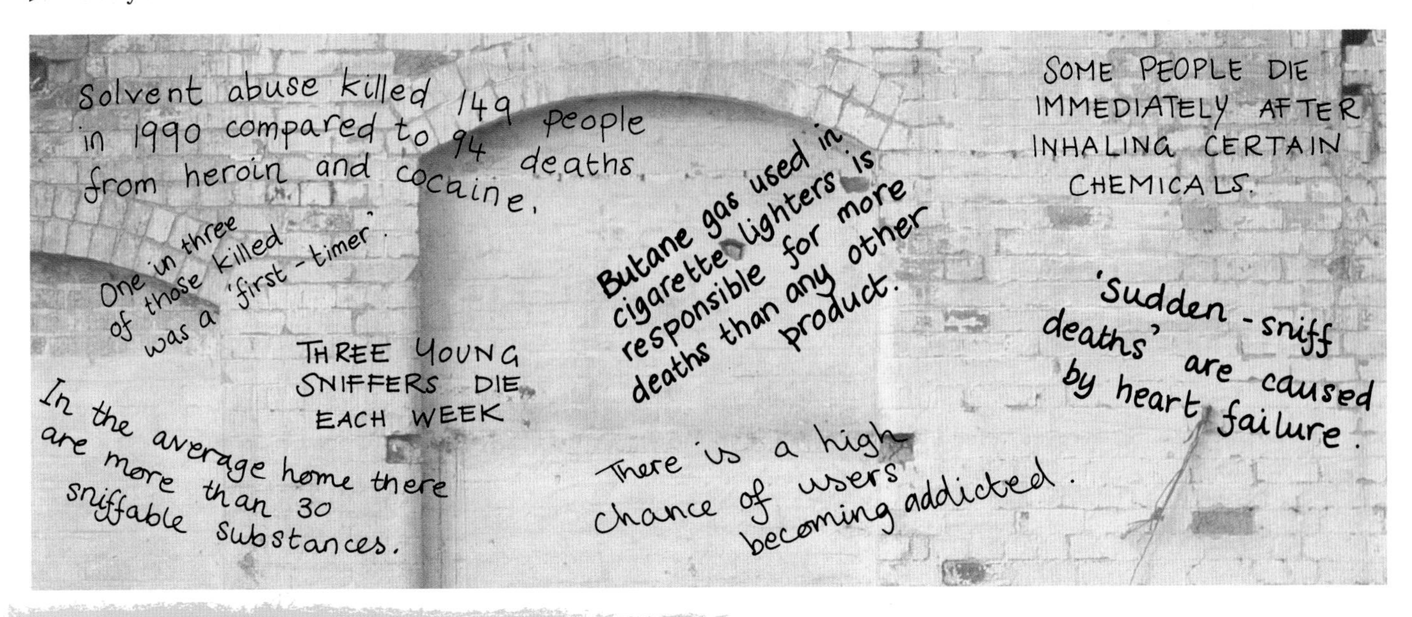

Lingering death of glue-sniffer Carl

A 21-year-old man was found hanged in a garage after 10 years of glue-sniffing.

When he was 11, Carl and some older friends went to a nearby building site and experimented with glue. He said "It was a laugh at the time and all the gang tried it."

His sniffing really started after he moved house and started at secondary school. His mother also had a new baby so he didn't get so much attention. Carl started to truant from school and discovered a gang of sniffers. At 14 he was taken into care – he felt rejected by his family. He shop-lifted to get glue and got into trouble with the police. By 18 he was often aggressive. He drifted from hostels to bedsits and often slept rough. What money he had was used to get more glue.

Carl aged 11 with friends

g Why do you think Carl started to sniff glue**?**

h What events in his life might have made him continue the habit**?**

i Why do you think that Carl truanted from school**?**

j How do you think people like Carl could be helped**?**

1 Copy and complete:
A drug is a . . . . that affects the . . . . system. Some . . . . save . . . . but others can kill if you take an . . . . Some people can't stop taking some drugs because they are . . . . Solvents are . . . . available and can attract . . . . people and those too . . . . to afford other drugs.

2 Explain how each of the following might make a few young people start to take a drug:
a) curiosity b) friends c) self-pity.
What type of person might be influenced most easily?

3 Two teenagers are found by a teacher, in an old house, inhaling solvent.
What happens next? What should the teacher do in the best interests of the teenagers and the school? Write the rest of the story.

4 Design a leaflet which shows some of the 'fundamental facts' about recreational drugs for the next Year 9 class to study this topic.

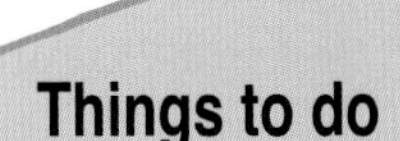

The shape you're in

Learn about:
- the skeleton
- how joints function

What keeps you in shape? In a word – **bones**.
But not all animals have bones.

a Can you name 3 animals that don't have any bones at all?

Larger animals need lots of bones to keep them in shape.
They need a **skeleton**.

b Write down some ways in which your skeleton helps you.

▶ Your teacher can give you a Help Sheet to find the bones in your body.
But you don't have to find them all – there are over 200!
Without your skeleton, you would feel really let down!

"I may be bony, but where would you be without me?"

Protection racket

The bones of your skeleton protect important organs in your body.

c Which part of your skeleton protects your brain?

d Which organs are protected by your ribs?

A great supporting act

Your main supporting bones are shaped like tubes.

- Balance a straw between 2 clamp stands.
- Measure the length of your straw.
- Hang a weight holder from the middle of your straw.
- Carefully add slotted weights until the straw collapses.
- Record the weight needed to bend the straw.
- Now repeat the test with half-length and quarter-length straws.

How does the length of the straw change its strength?

- Now try to extend your investigation by changing the ***number*** of straws that you use.

Joints

▶ Try walking without bending your knees, keeping your legs quite straight.
It's not very easy is it?
So how do you think we move our bones?
Bones can move at **joints**.

Look at these different types of joints:

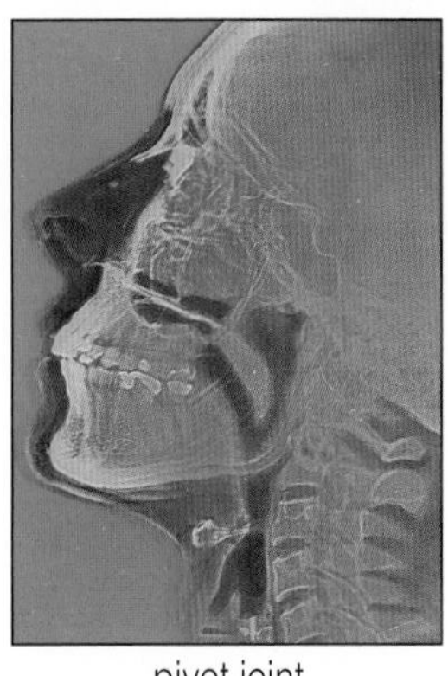
pivot joint

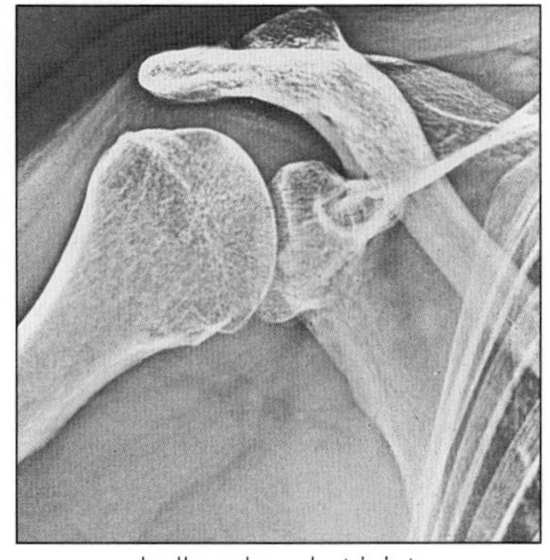
ball and socket joint

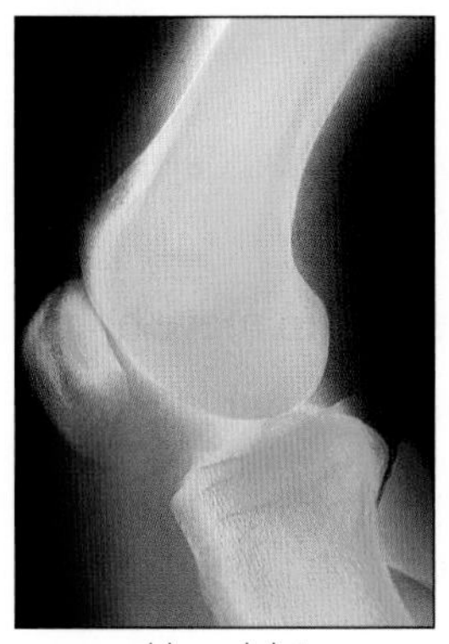
hinge joint

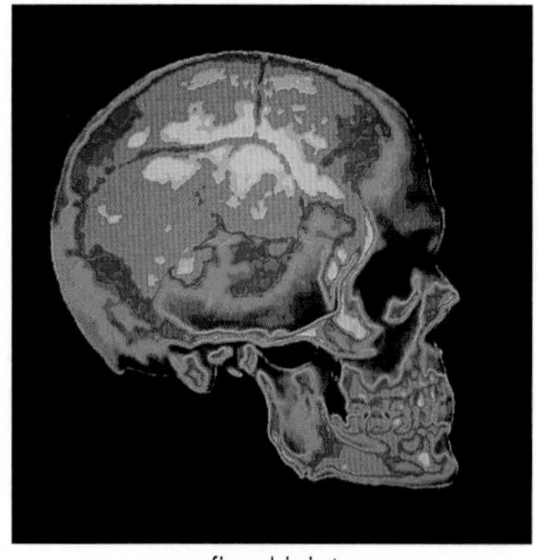
fixed joint

Where do you think they are found in your body?
What type of movement can you make at these joints?

▶ Copy and complete this table with your answers:

Type of joint	*Where found*	*Type of movement*
pivot ball and socket hinge fixed	neck	nodding or turning

Things to do

1 Copy and complete:
The . . . . in your skeleton protect many important . . . . in your body.
Bones also allow . . . . to occur at joints.
Your elbow is an example of a . . . . joint and your shoulder is an example of a . . . . and . . . . joint. Many bones are shaped like a tube. This is a good shape to . . . . your body.

2 Use your Help Sheet to find:
a) the largest bone in your body, and
b) the smallest, bone in your body.
On your Help Sheet:
c) label as many bones as you can, and
d) shade in red the bones that protect important organs.

3 Find out as much as you can about what these do:
a) **tendons** b) **ligaments** c) **cartilage**.

4 X-rays can penetrate through skin and soft tissue but not so easily through bones. X-rays can help doctors to find out if bones are broken.
a) Which bones are fractured in this X-ray?
b) How do you think the patient will be treated to repair the broken bones?
c) When an X-ray is taken of the gut, the patient drinks a liquid that will not let X-rays through. Why do you think this is?

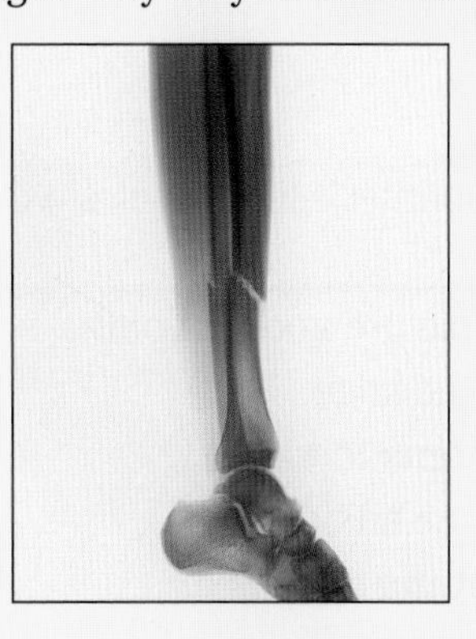

Moving muscles

Learn about:
- how muscles work

Are you as muscular as the woman in the photograph? Probably not. But you still have over 350 muscles that do important jobs.

a How do you think the body-builder has developed such powerful muscles?

Your muscles provide the force needed to move bones at **joints**.

▶ Feel your calf muscle at the back of your leg.
Now lift your heel but keep your toes on the floor.
Can you feel your muscle pulling? What happens if you over-exercise a muscle that you don't use much normally?

No pushing!

Muscles cannot **push** – they can only **pull**.

Push up against the underside of the bench with the front of your hand.

b What does your **biceps muscle** feel like?

When it pulls it gets shorter and fatter – we say that it **contracts**.

Now push down with the back of your hand against the bench-top.

c What does your **triceps muscle** feel like?

Your triceps contracts to pull your arm straight.

When a muscle is not contracting, it returns to normal size.
We say that it **relaxes**.

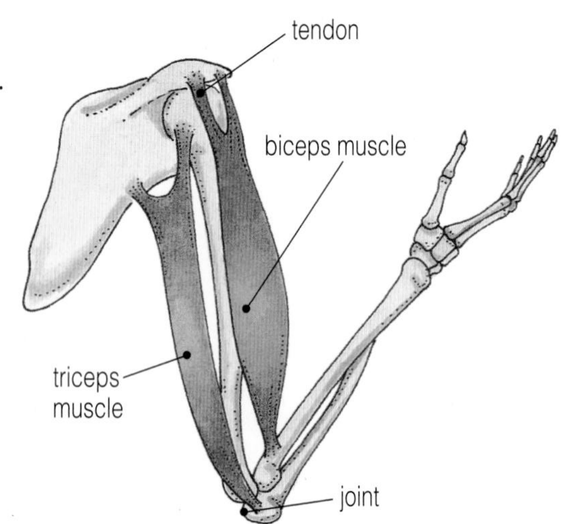

Muscles like your biceps and triceps work in pairs.
When one contracts the other relaxes.
We say that they are **antagonistic**.

▶ Copy and fill in the table to show which muscle contracts and which relaxes:

	Biceps muscle	*Triceps muscle*
Pushing up with the front of your hand		
Pushing down with the back of your hand		

Muscles at work

The pictures show how a sprinter's leg muscles work at the start of a race.

▶ Look carefully at each picture and find out:

d Which muscle **bends** the knee.

e Which muscle **straightens** the knee.

f Which muscle **bends** the ankle.

g Which muscle **straightens** the ankle.

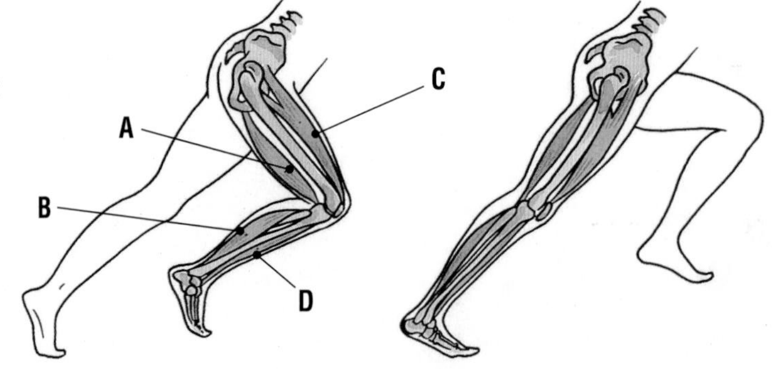

Mighty movers

You can test your finger strength in this investigation.

- Arrange the clamp stand as shown.
- Place your hand flat on the table.
- Put your middle finger through the rubber band.
- Now move your finger down to touch the table.
- Count the number of these finger movements you can do continuously for 2 minutes.
  Be sure to touch the table each time and keep your hand flat.
- Record the number of finger movements for each 20-second period in a table like this:

Time interval (seconds)	0–20	20–40	40–60	60–80	80–100	100–120
Number of finger movements						

h Plot a line-graph with axes like this:

i What sort of pattern was there to your results?

j Try to explain any pattern that you observed.

k Describe how you feel after doing the exercises. Why do you feel like this?

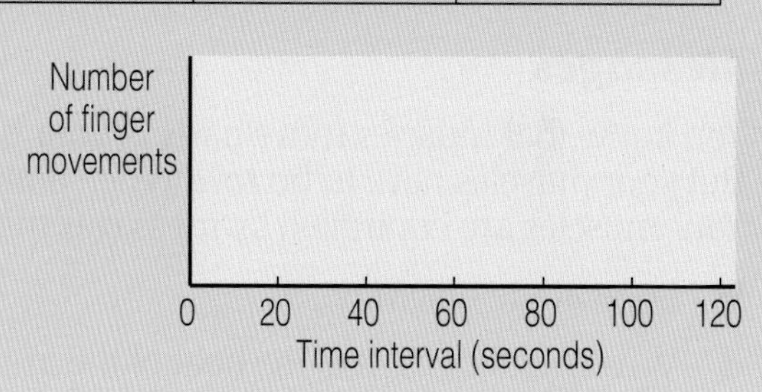

1 Copy and complete:
Your . . . . provide the force to move . . . . at joints.
A muscle cannot . . . . ; it can only . . . .
When a muscle pulls it gets . . . . and fat.
We say that it . . . .
When a muscle is not contracting we say that it . . . .

2 Plan an investigation to find out whether exercise or diet is more important in increasing muscle size.

3 Make a model of an arm using your Help Sheet.
Glue the sheet onto cardboard and cut out the shapes of the bones. Join them together with a paper fastener. Use elastic bands for the muscles.
How much weight will your model support?
Try to evaluate how much your model works like the real thing.

Things to do

Moving parts

Learn about:
- nerves and reflexes
- friction and joints
- medical research

What happens if you accidently touch a hot iron?
If you have any sense, you move away!

a How quickly do you move away?

b Why do you think that you move away quickly?

c How do you think that this happens?

You pull your hand away so quickly because messages are sent around your nervous system at high speed. These tell you what is happening and what to do. This is an automatic action because you do it without thinking.

An ***automatic*** action like this is called a **reflex**.

Messages

You know that muscles move parts of your body.
But your muscles have to be ***told*** when and how to work.
Your muscles are controlled by messages that travel along **nerves**.

▶ Look at the diagram:

spinal cord and brain

muscle

heat sensors in skin

d What is it that detects the heat of the iron?

e Our skin is a **sense organ**.
Do you know any other sense organs in your body?

f What do you think happens when the messages reach the brain and spinal cord?

g What happens when the messages reach the muscle?

Reducing the friction

Your bones move a lot at **synovial joints**.
At these joints, there are tough **ligaments**.
The ends of the bones have a layer of **cartilage**.
Synovial fluid covers the surface of the cartilage.

▶ Look at this diagram of the hip joint:

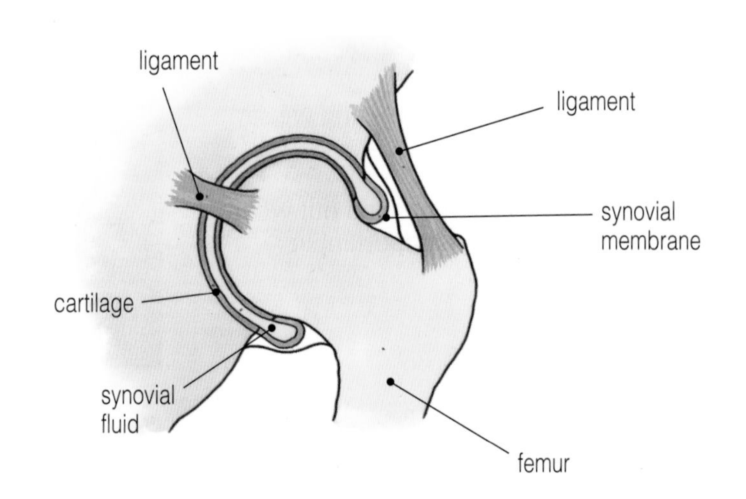

Which part of this synovial joint do you think:

h holds the bones together?

i reduces friction?

j holds the synovial fluid in place?

k acts as a shock absorber?

Damaged cartilage is a common **sports injury**.
Footballers can have operations to remove damaged cartilage from around the knee. Later in life, many retired footballers need replacement hip joints.

▶ Draw a flow-chart that shows the different scientists that an ex-footballer can thank for his new hip joint.

There's a catch

How fast are your reactions?
Do you think they speed up with practice? Do drugs affect them?
Caffeine is a stimulant found in tea, coffee and cola.
Plan an investigation to find out if caffeine in cola affects your reaction time.

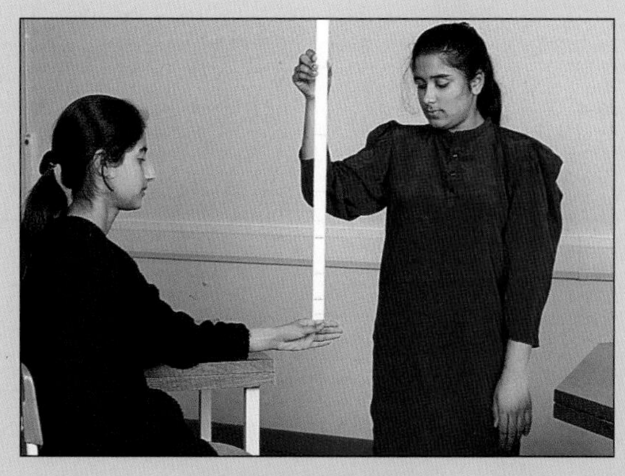

You can measure your reaction time with a falling ruler, as shown in the photograph:

- Drink a known volume of de-caffeinated cola.
- Then add a time scale to the ruler.
  Your teacher will give you a Help Sheet for this.
- Then place your arm on the bench as shown:
  Your partner holds the ruler with the zero next to your little finger, but ***not*** touching it.
  When your partner lets go of the ruler, try to catch it as quickly as possible.
  Read the scale next to your little finger.
  This shows how long you took.

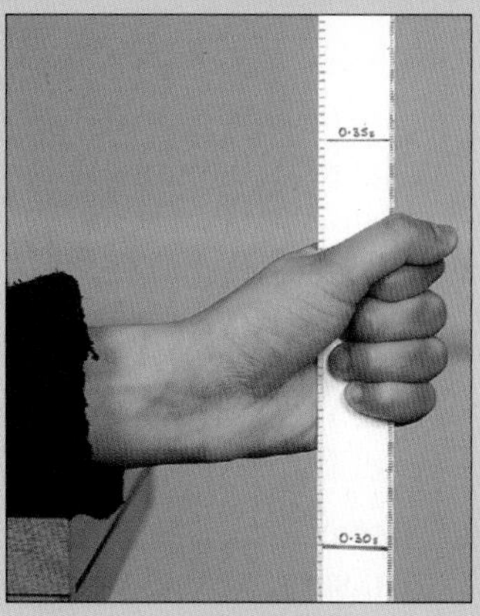

- Now repeat this test 10 times.
  You can record your times in a table like this:

Trial	Time (seconds)	
	Without caffeine	With caffeine
1		
2		

- You can now repeat the test once you have drunk the same volume of caffeinated cola.

l Was your reaction time quicker with or without caffeine? Why do you think this is?

m Which variables could you control in this enquiry?

n We know that each individual differs. How could you gather more reliable evidence?

In **medical research**, scientists have to carry out trials to see if new drugs are effective (as well as assessing their side effects).
During their trials,

- why are some volunteers given **placebos** that do not contain the new drug?
- why are scientists judging the effects on volunteers who are not told who has taken the placebo and who has taken the new drug?

Things to do

1 Copy and complete:
Sense . . . . in our bodies sends messages at . . . . speed through our . . . .
The messages get our . . . . to move our bodies.
When our nerves work in this way it is called a . . . . action.
Our reflexes are automatic. They work very . . . . and often . . . . us from harm.

2 a) What is meant by your reaction time?
b) Name 3 sports in which you think that reaction time is:
i) important ii) not important.
c) If you were a tennis player, how could you try to improve your reaction time?

3 Which animals have quick reflexes?
Give some examples of situations where animals need quick reactions in nature.

4 How do you think each of the following can affect a person's reaction time:
a) tiredness? b) coffee (caffeine)?
c) alcohol? d) practice?

Biology at Work

Learn about:
- sports injuries
- arthritis
- coronary heart disease

Sports injuries

In some sports a person runs the risk of being injured.

a What parts of the body have the most common injuries?

Sprains occur in ligaments.

b What do ligaments join together?

c What do you think happens to a ligament when a sprain occurs?

d How would you treat a sportsperson with a sprain?

If ligaments tear badly, then a dislocation can occur at a joint.

e What is a dislocation?

f What is meant by a pulled muscle?

Fractures are broken bones.
In which sports could the following fractures occur:

g fracture of the radius (the lower arm)?
h fracture of the collar bone?
i fracture of the ankle?

j Why do you think many injuries occur *early* on in a sports season?

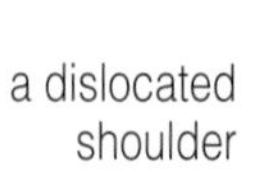

a dislocated shoulder

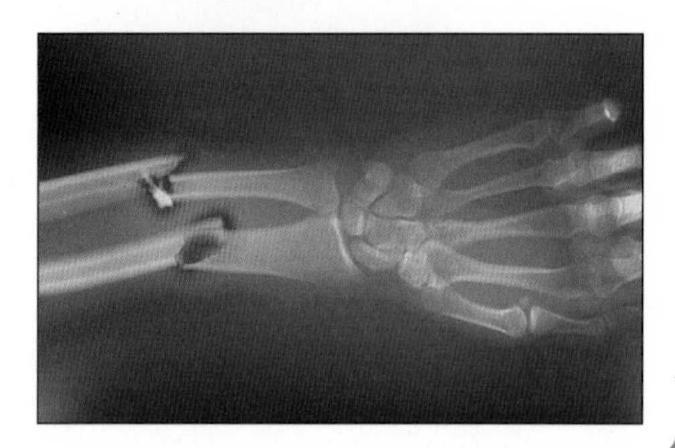

Arthritis and artificial hips

As we get older our joints don't work as smoothly.
Friction at a joint can cause pain.

k Which fluid cuts down friction at a joint?

l Which tissue acts as a shock absorber at a joint?

Osteoarthritis affects older people.

m Use the diagram to explain how osteoarthritis occurs.

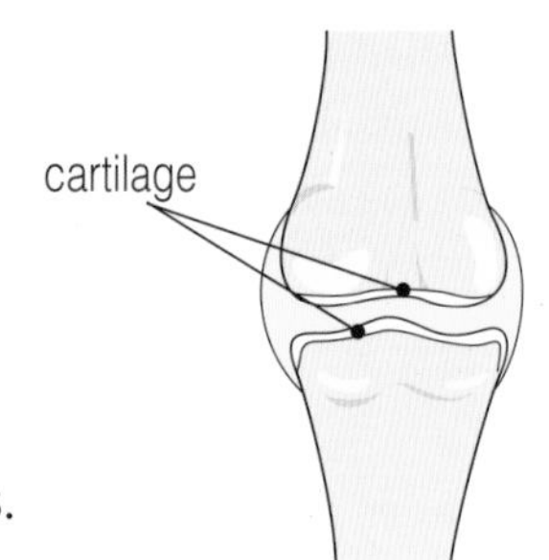

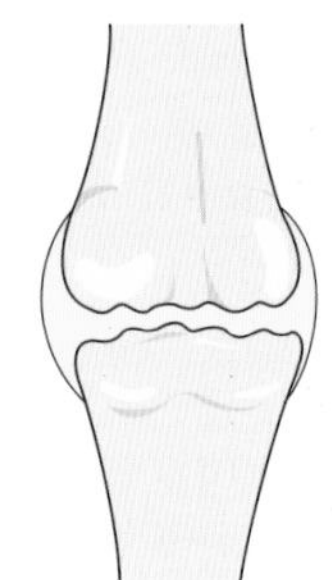

Rheumatoid arthritis is an inherited condition.

n Find out how rheumatoid arthritis can affect a joint.

Arthritis can make it very painful for people to walk.
Some people have operations to replace their hip joints with artificial ones like the one in the picture:

o Why is the hip joint a 'ball-and-socket' joint?

An artificial hip joint must have the same properties as the natural one.

p Write down what you think these properties are.

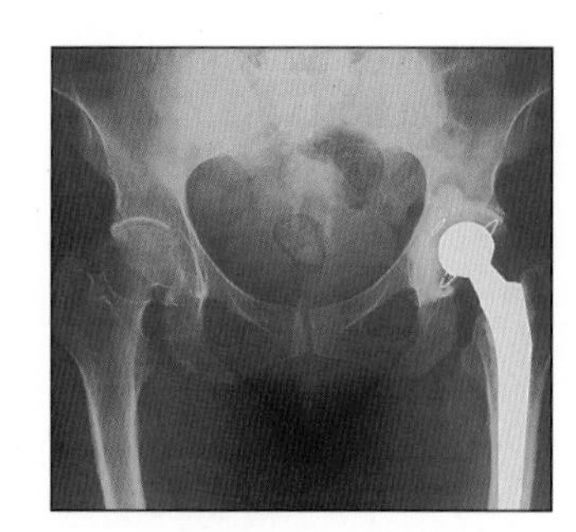

Blood pressure

When the heart pumps it produces a pressure in your arteries.
We call this **blood pressure**.
Your blood pressure rises if you do anything to make your heart beat faster or if your arteries become narrower.

q What might cause your heart to beat faster?

r What could cause your arteries to narrow?

A stroke victim 'learning' to walk again

Constant high blood pressure is harmful.
It puts a strain on the heart and makes it work harder.
It can also cause an artery to burst.
If this happens in the brain it can cause a **stroke**.

s Why would a stroke cause damage to the brain?

t What sort of effects can a stroke have on a person's body?

If an artery bursts in the heart it can cause a **coronary heart attack**.

u Why would this cause damage to the heart muscle?

v What sort of things could increase the chances of getting coronary heart disease (hint: think of diet, exercise and lifestyle).

w Find out how a **coronary artery by-pass** can relieve patients of the symptoms of coronary heart disease (**CHD**).

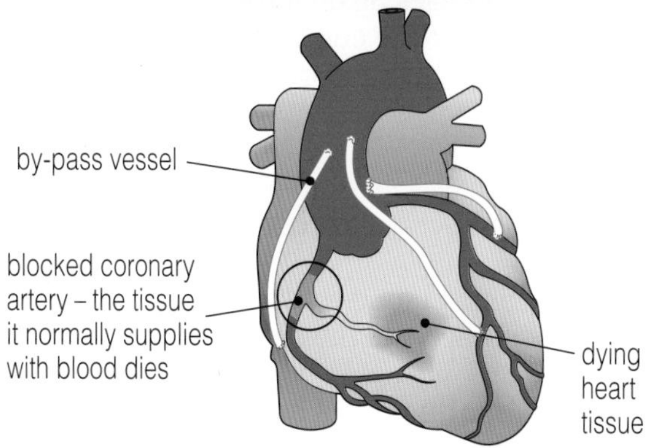

The right exercise

People can exercise to reduce the effects of CHD.
But the exercise must be *appropriate* to a person's age, weight and lifestyle.

If you are fit, you will have a **low heart rate** and **quick recovery time** after exercise.

Draw a bar-chart of the heart rates of the 5 people in the table:

x Explain why the heart rates are different for each person.

The table shows the heart rates of different people.

People	*Heartbeats per minute*
Fit teenager at rest	65
Fit teenager, exercising	170
Unfit teenager, exercising	190
Fit adult at rest	70
Unfit adult, exercising	200

Draw line-graphs of the recovery rates for the fit and unfit teenagers after exercise shown in this table:

y Compare the recovery rates of the two teenagers after exercise.

z Use your graphs to explain what sort of people have fast recovery rates.

	Heart rate per minute						
	0	1	2	3	4	5	6
Unfit teenager	190	160	135	115	103	95	88
Fit teenager	170	120	90	80	72	65	65

▶ Consider this question: 'Are we healthier than our great-grandparents were?'

As you know, there are different aspects of a healthy life style.
Think of a question related to the one above.
Carry out research to gather evidence to answer your question.
Share your findings with other groups and discuss the strength of evidence for each point of view.

The active body

Learn about:
- how ideas about the body changed over time

Ideas about blood circulation

Some of the first ideas about the circulation of blood were put forward in the second century AD by the Roman physician **Galen.**

He thought that blood seeped out of the heart to various parts of the body and back again through the same blood vessels. Galen likened it to 'the tide flowing in and out of an estuary'. The dissection of human bodies was not allowed in Rome in Galen's time.

Ibn Al-Naphis was an Arab physician of the thirteenth century. He was the first person to connect the functions of the heart and lungs.
He suggested that blood was purified in the lungs where it was refined on contact with the air inhaled from the outer atmosphere.

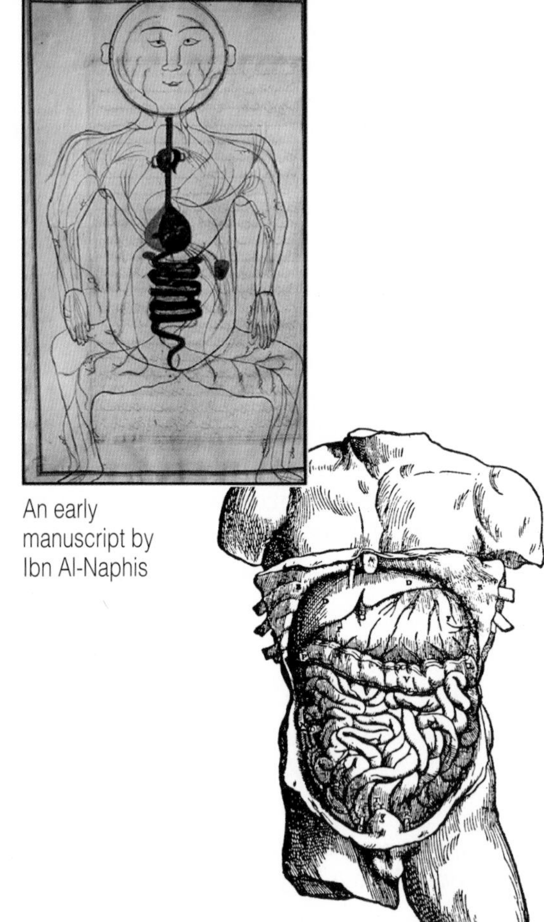

An early manuscript by Ibn Al-Naphis

An illustration of a dissection by Versalius

Andreas Versalius was a Belgian, teaching in the University of Padua. He encouraged his students to dissect human bodies.
In 1543, Versalius published a book called *On the Fabric of The Human Body.* It contained accurate and detailed accounts of the organs and structure of the human body. It had superb illustrations based upon dissections that Versalius had carried out. It was the first detailed book on **anatomy**.

William Harvey was an English physician who used this anatomical knowledge, and respect for detail, to make a unique discovery of the circulation of blood around the body.
In 1628, he published his book *On the Motions of the Heart and Blood*.
In it Harvey explained that blood moves around the body in a circle, from the left side of the heart in arteries and then back to the right side in veins. He also suggested that there are thousands of tiny blood capillaries connecting arteries to veins.
But he was never able to see these connections between arteries and veins as microscopes hadn't been invented.

Harvey demonstrating an experiment to a group of physicians in London

Harvey's experiments

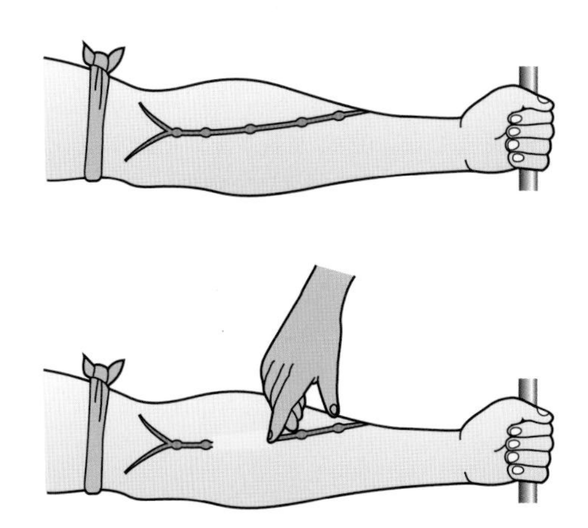

Harvey based a lot of his ideas upon ***detailed observations*** and clever experiments, using ***deductive reasoning***.
The evidence collected from observations is called ***qualitative data***.

One of these experiments is shown in the diagram and demonstrates the flow of blood towards the heart in veins.

It shows that when the arm is lightly tied with a bandage, the veins show up with small swellings where the valves are.
Blood can be pushed past these points towards the heart but it cannot be pushed back.

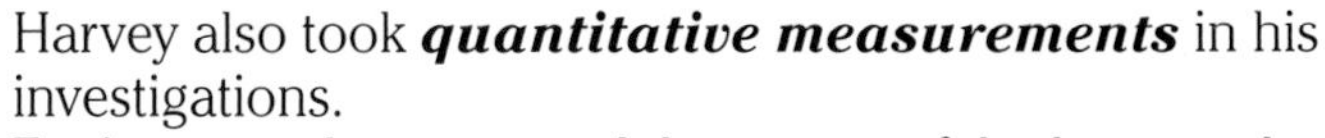

Harvey also took ***quantitative measurements*** in his investigations.
For instance, he measured the output of the heart and was able to calculate the amount of blood in the circulatory system in any given time.

Harvey was the first person to study the **physiology** of the human body, that is, how the body works.

Around this time there was a great deal of interest in anatomy and physiology: scientific societies were established, schools were founded and books published.

Towards the end of the seventeenth century, the Italian physiologist **Marcello Malpighi** was to prove Harvey right when he observed and described capillaries.

He used a microscope to examine the webbed foot of a frog and was able to see blood flowing through capillaries.

They did indeed form the link between arteries and veins and were thin enough for substances to be exchanged between the blood and the cells.

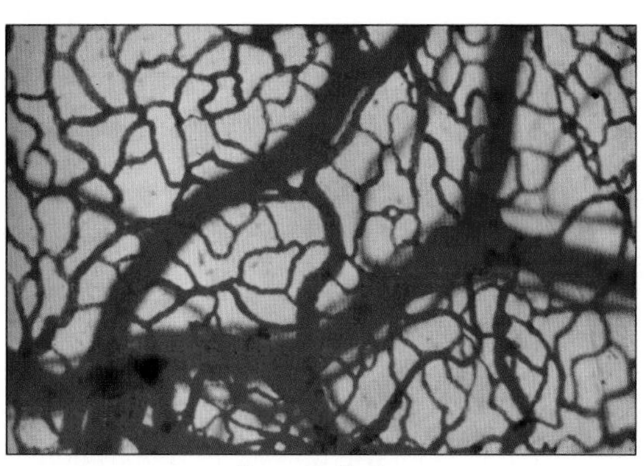

Frog capillaries

Questions

1 a) What do we mean by the study of i) anatomy and ii) physiology?
b) Why do you think a person needs to study anatomy before he or she can understand physiology?

2 Draw a simple diagram to show Harvey's ideas about the circulation of blood in the human body. Use the following words: veins, arteries, right and left sides of the heart and capillaries.

3 Scientists work in many different ways. Use the account above to list as many ways as you can.

4 Try to explain what is meant by the following phrases in the account:
a) detailed observations
b) deductive reasoning
c) quantitative measurements.

5 Try out Harvey's experiment on veins with a friend. Make sure that you have your teacher's permission and do not tie the bandage too tight.

6 Choose one of the scientists and find out more details about his life.

Questions

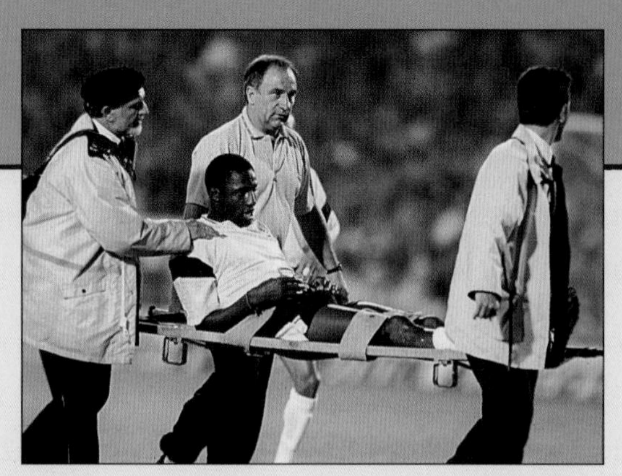

1 Many sports injuries affect muscles and bones.
Use a first aid book to find out as much as you can about:
a) fractures b) torn muscles c) dislocations.
Which sports do you think are more likely to cause each type of injury?

2 Devise a programme of exercise to develop:
a) the biceps,
b) the triceps, and
c) the pectoral muscles.

3 a) What joins muscles to bones?
b) What joins bones to bones?
c) What cuts down friction at a joint?
d) What is arthritis?

4 A young child runs out into the road from between 2 parked cars.
A car driver reacts very quickly by slamming on the brakes.
The car screeches to a halt. Luckily the child is unhurt.
Explain, as fully as you can, the way in which the driver's nervous system worked.

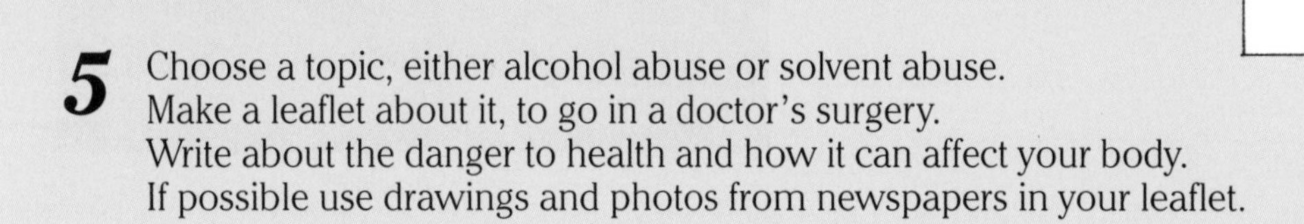

5 Choose a topic, either alcohol abuse or solvent abuse.
Make a leaflet about it, to go in a doctor's surgery.
Write about the danger to health and how it can affect your body.
If possible use drawings and photos from newspapers in your leaflet.

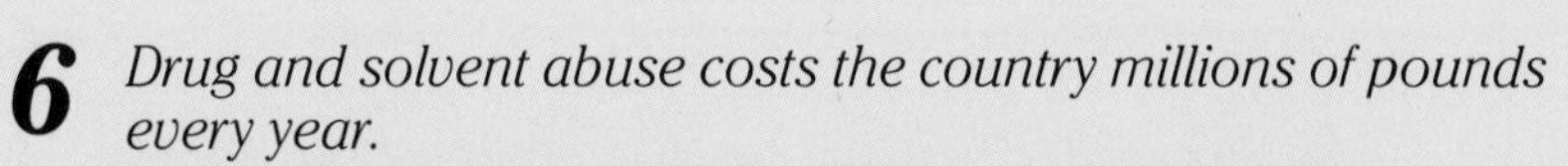

6 *Drug and solvent abuse costs the country millions of pounds every year.*
Use the following headings to help you to explain this sentence: a) medical treatment b) burglary and theft c) catching the drug pushers.
What actions do you think would reduce drug abuse?

7 Suppose that you have a young kitten or puppy.
Plan an investigation to measure its growth.
How will you measure its growth (try 2 ways)?
How often will you take measurements?
How long do you think this investigation will take?
How will you display your results?

Plants and photosynthesis

Can you imagine what life would be like without plants?

We use plants for food, fuel, building materials and medicines. Plants also take waste carbon dioxide out of the air and make oxygen for us to breathe, thanks to the process we call photosynthesis.

In this unit:

The energy trap

Learn about:
- photosynthesis and biomass
- testing leaves for starch

Plants can't eat like animals do.
So how do they get their food?

▶ Write down some ideas about how you think plants feed.

Plants make their food from simple substances.
But to do this they need energy.
Where do you think this energy comes from?
What part of the plant do you think traps this energy?

Plants make food by the process **photosynthesis** (photo = light *and* synthesis = to make). To make food they need:

- carbon dioxide from the air
- water from the soil
- light energy trapped by **chlorophyll**.

CARBON DIOXIDE + WATER —SUNLIGHT / CHLOROPHYLL→ SUGAR (GLUCOSE) + OXYGEN

▶ Write down the answers to these questions:

a What food do plants make themselves?

b Which gas is made during photosynthesis?

c How do animals use this gas?

Oxygen bubbles

Jill and Emma had seen fish tanks with air bubblers. They knew that these were to put oxygen into the water for the fish to breathe. Jill read that if there is lots of pondweed and the tank is well lit, air bubblers aren't needed.

Jill and Emma set up an investigation to see the effect of light on the pondweed.
They put some pondweed in a test-tube of pond water.
They placed a lamp at different distances from the test-tube and counted the number of bubbles of gas produced in a minute.
Their results are shown in the table:

Distance of lamp from pondweed (cm)	10	20	40	Lamp off
Number of bubbles produced per minute	15	7	4	2

d Which gas do you think was produced by the pondweed?

e How could you prove this?

f Do you think there is a pattern to these results?

g Emma thought that the lamp might also warm up the pondweed and not make it a fair test. How could you improve the design of their investigation to avoid this?

▶ If you have time, try out the investigation on pondweed yourself. Write a report, including tables and graphs to display your results. Identify any anomalous results.

Testing a leaf for starch

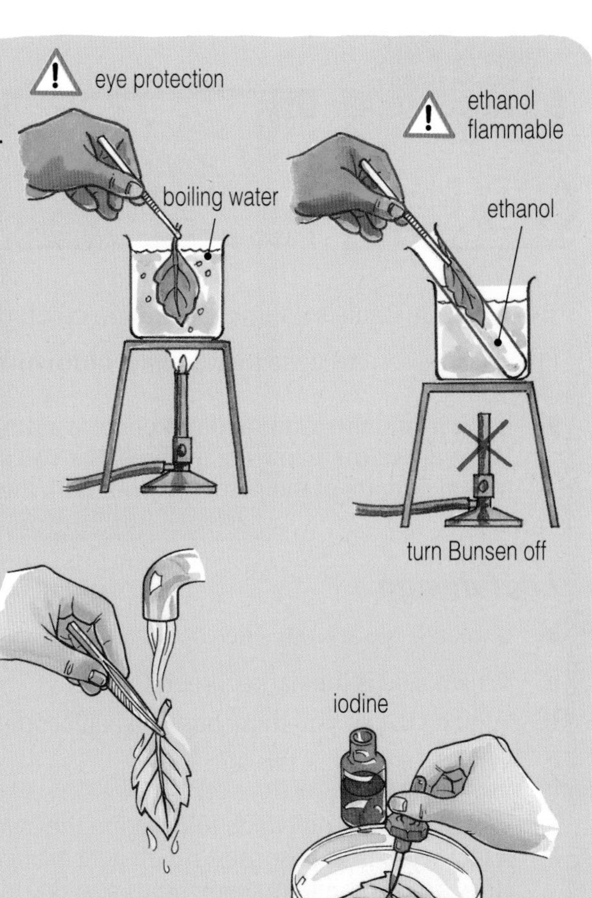

Most of the sugar made in the leaves of a plant is changed to starch.
You can test for this starch with iodine.
If the leaf turns blue-black with iodine then starch has been made.

- Dip a leaf into boiling water for about 1 minute to soften it.
- Turn off the Bunsen burner.
- Put the leaf into a test-tube of ethanol. Stand the test-tube in the hot water for about 10 minutes.
- Wash the leaf in cold water.
- Spread the leaf out flat on a petri dish and cover it with iodine. What colour does the leaf go?

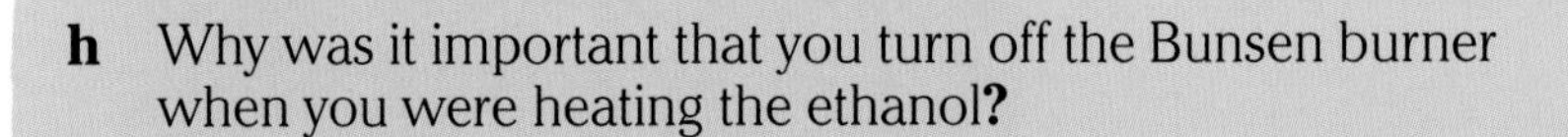

h Why was it important that you turn off the Bunsen burner when you were heating the ethanol?

i What was the leaf like after you heated it in the ethanol?

j Was there any starch in the leaf that you tested?

k Plants make new **biomass** during photosynthesis. What is meant by biomass? (Hint: see 7I4 in *Spotlight Science 7*.)

Stripes and patches

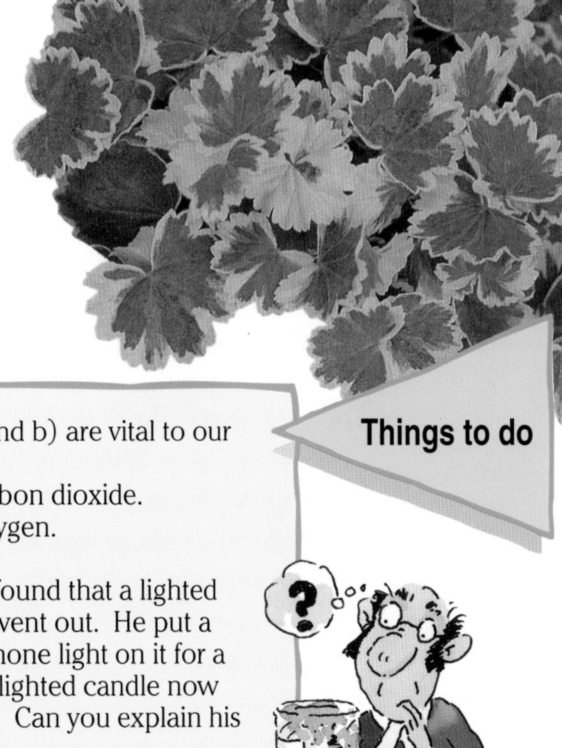

Not all leaves are green all over.
Some have white and green patches, others stripes.
If you have time, try testing one of these leaves for starch.

l Which parts do you think will go blue-black?

Remember to draw a picture of what your leaf looks like at the start to show which parts are green.

Things to do

1 Copy and complete:
Plants make their food by . . . . They use . . . . from the air and . . . . from the soil. They also need a green substance called . . . . which traps . . . . energy. The food that is made is sugar and it is changed to . . . . in the leaf. The waste gas made is called . . . .

2 Plants are important because they provide:
a) food b) fuel
c) building materials d) medicines.
Find examples of plants that provide each of these things.

3 Explain why a) and b) are vital to our survival.
a) Plants use up carbon dioxide.
b) Plants release oxygen.

4 Joseph Priestley found that a lighted candle in a jar soon went out. He put a plant in the jar and shone light on it for a week. He found the lighted candle now burned much longer. Can you explain his experiment?

Leaves

Learn about:
- the structure of leaves
- respiration in plants

What do you think is the most common colour in nature?

Plants are green because they contain **chlorophyll**.

▶ Look round the room and out of the window.
Write down the names of 5 plants that you see.
In which parts of the plants do you think there is most chlorophyll?

Leaf design

▶ Look closely at both sides of a leaf.

a Which side is the darker green?

b Which side do you think has most chlorophyll?

c Why do you think this is?

▶ Write down some words to describe the shape of your leaf.
A leaf's job is to absorb as much light as it can.
How do you think its shape helps it to do this?

Looking inside a leaf

Look at a section of a leaf under your microscope.
Focus onto the leaf at low power.
Now carefully change the magnification to high power.
Can you see any of the parts labelled in the diagram on your slide?

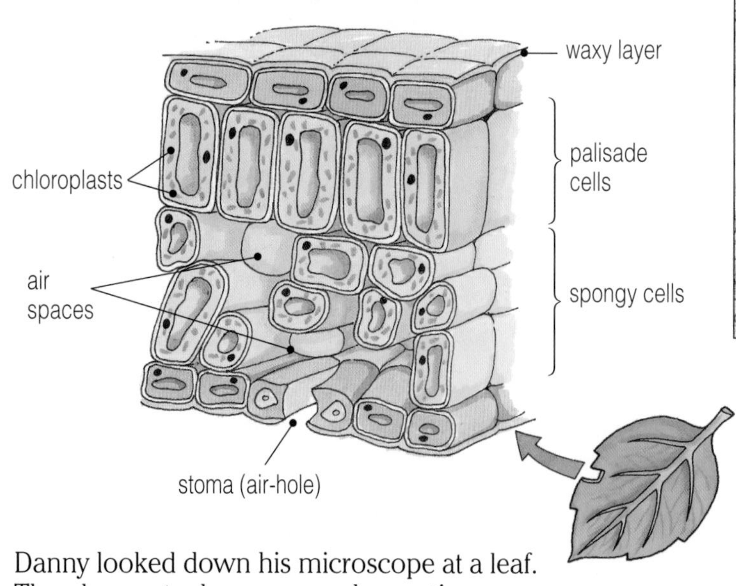

Danny looked down his microscope at a leaf.
Then he wrote down some observations:

1. Palisade cells contain lots of chloroplasts.
2. Palisade cells are in the upper half of the leaf.
3. A waxy layer is on the upper surface.
4. Lots of air-holes on the lower surface.
5. Lots of air spaces between the spongy cells.

▶ Look at each of Danny's observations:
Say how you think each adaptation helps the leaf to do its job.

Holey leaves

Try dropping a leaf into a beaker of boiling water.
Which surface do bubbles appear from?
Why do you think this is?

⚠ eye protection

Gases from the air pass into and out of a leaf through **stomata** (air-holes).

Paint a small square (1 cm × 1 cm) on the underside of a leaf with nail varnish.
The nail varnish will make an imprint of the leaf surface.

Wait for it to dry completely.
(While you are waiting, you can set up your microscope.)
Carefully peel off the nail varnish with some tweezers.
Put it on a slide with a drop of water and a coverslip.
Observe and draw 2 or 3 stomata at high power.
Repeat for the upper surface of your leaf.

Write down your conclusions.

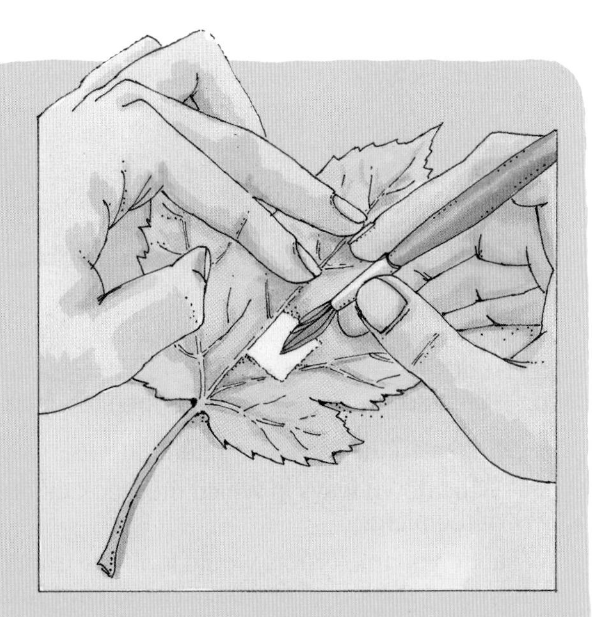

In and out of leaves

Two very important living processes take place in plants.
Can you remember what they are?
They both involve gases passing in and out of the leaves through stomata (air-holes).

d Look at the diagram **A**:
 i) Which gas passes into a leaf during the day?
 ii) Does photosynthesis take place faster or slower than respiration, during the day?

e Look at the diagram **B**:
 i) Which gas passes out of a leaf at night?
 ii) Does photosynthesis or respiration take place ***all the time*** in green plants?

The glucose from photosynthesis is also used by plants to build new biomass. For example, glucose molecules join together to form starch (an energy store) and cellulose (part of the plant cell wall).

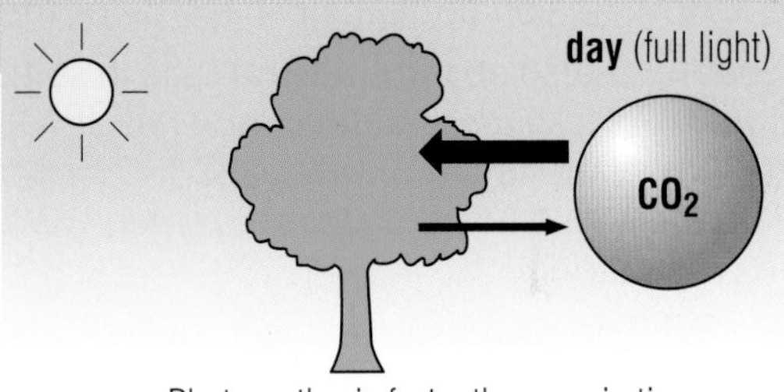

A Photosynthesis faster than respiration. Plant takes in more carbon dioxide than it gives out.

night (dark)

CO_2

B Photosynthesis stops, respiration continues. Plant gives out carbon dioxide but no longer takes it in.

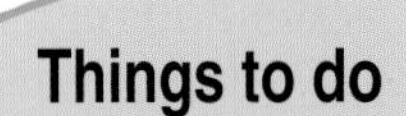

1 Match each of the leaf parts on the left with the job that they do on the right:

stomata	carry water up from stem
palisade cells	allow gases to pass into and out of leaf
spongy cells	contain many chloroplasts
waxy layer	contain many air spaces
veins	prevents too much water being lost

2 Do plants of the same species always have the same leaf area?
Plan an investigation to compare the leaf area of the same plants found in light and shady conditions. Try to explain any differences that you find.

3 Leaves have a large surface area to absorb light. Lay a leaf onto graph paper, draw around it and work out its area by counting the squares. Can you work out the total leaf area of the plant?

4 Leaves are thin so gases can get in and out easily. But they can also lose water and then they droop.
Look closely at a leaf and say what helps to stop them from drooping.

5 Draw a poster showing the variety of ways we use biomass. You will need to use secondary sources to find your information.

Plant plumbing

Learn about:
- the function of roots
- the importance of water

The roots of a plant grow into the soil.

a Write down what you think the roots are for.
b What do you think would happen to a plant if it had no roots?

▶ Look carefully at some different roots.

c Write down ways in which they look different from the rest of the plant.

What do you think roots would look like under the microscope?
You would see lots of tiny hairs called **root hairs**.
The root hairs take up water from the soil.

d How do you think their shape helps them to do this?

Root hairs also absorb mineral salts in solution from the soil.
If a plant lacks any essential minerals, it will not grow into a healthy plant. (See 9D2, page 52.)
The roots also need to absorb oxygen gas from the soil to use in respiration.

e Why do plants die in water-logged soil even though they can get plenty of water and minerals?

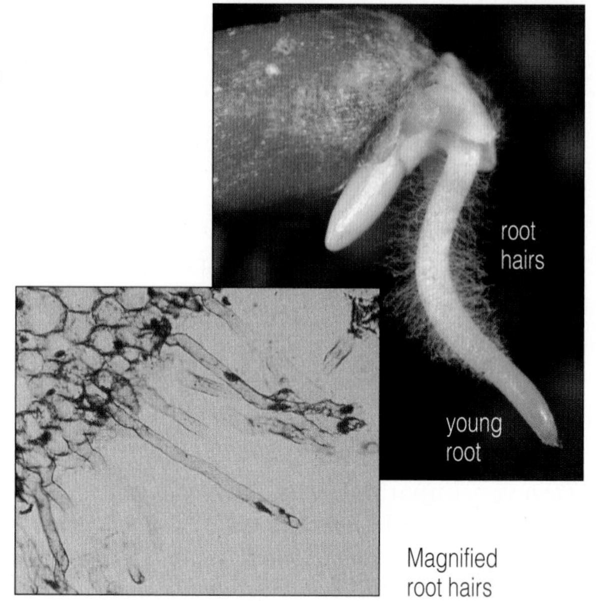

Magnified root hairs

Up to the leaves

Your teacher will give you a piece of celery.
This has been standing in water containing a dye.

Carefully cut off about 1 cm as shown in the photograph.
Make an accurate drawing of the inside of the celery.
Colour the parts where you can see the dye.

The dyed water is carried in tiny tubes called **xylem** (sigh-lem).

Carefully cut out a 2 cm length of your xylem.
Look at it with a hand-lens. Describe what you see.

Stem support

Water and mineral salts pass into a plant through the roots.
They then pass in the xylem up the stem to the leaves.

f What else do you think the stem does?
g What would happen to the leaves and flowers if there was no stem?

There are other tiny tubes in the stem called **phloem** (flow-em). Look at the diagram:

h Write down what you think the phloem does.

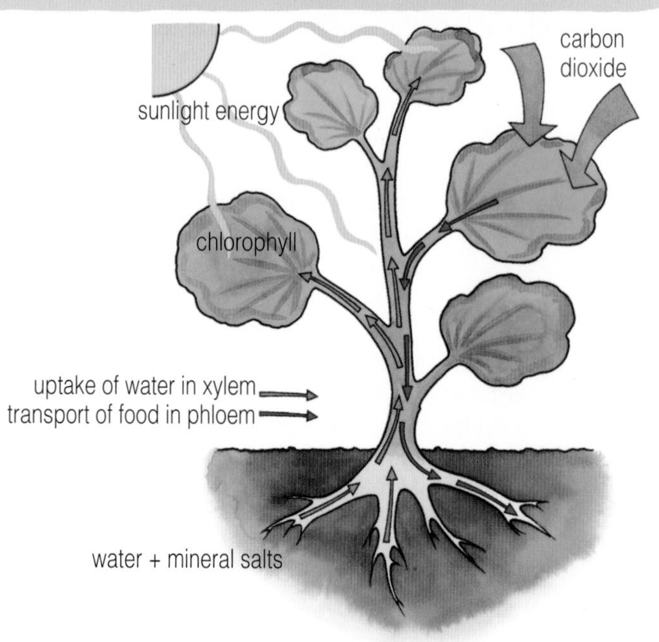

Waterways

How do you think water travels up to the tops of tall trees?
As water is lost from the leaves, more water moves up from the stem.

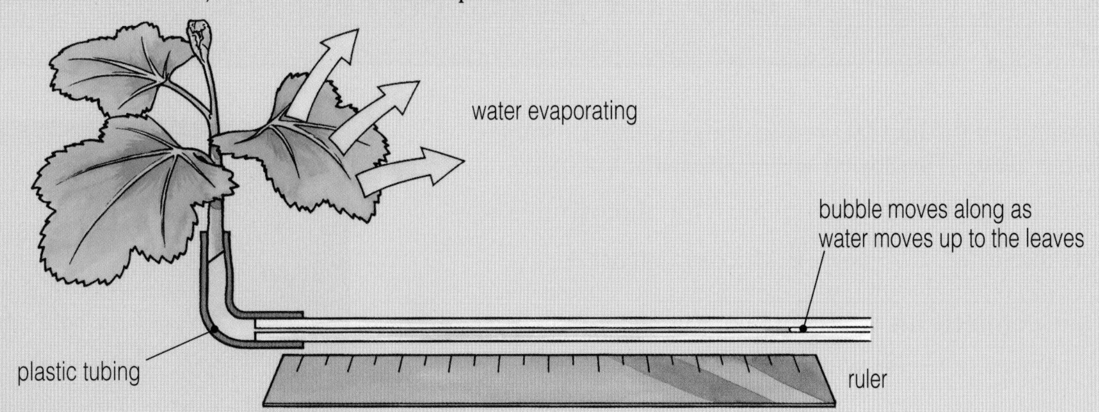

You can measure how quickly water is moving up the stem to the leaves with this apparatus. Your teacher will show you how to set it up.

Measure the distance travelled by the bubble every minute.

Plot your results on a line-graph with axes like this:

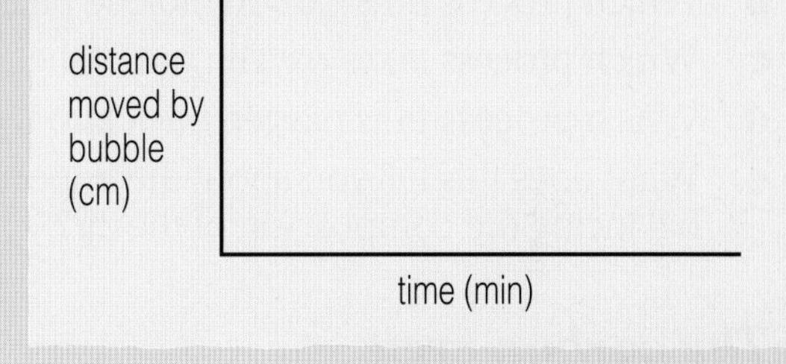

This experiment was repeated by 2 groups of pupils:

- Group A covered the shoot with a clear polythene bag
- Group B placed a fan close to their apparatus.

Here are their results:

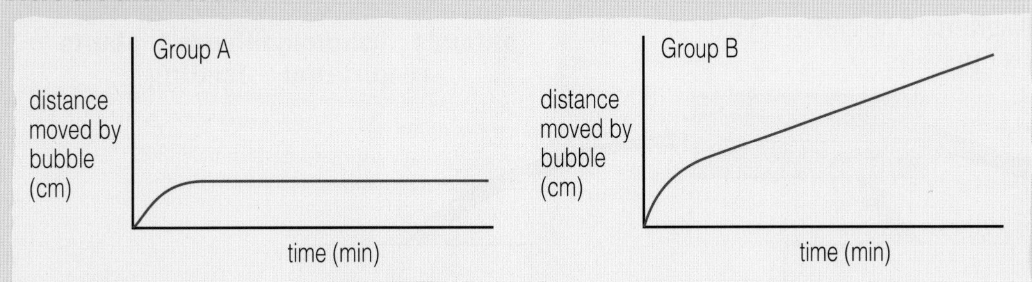

Plants need water to:

- use in photosynthesis
- transport minerals
- keep firm
- help cell growth
- cool leaves down
- form fruits.

i What conditions are being tested by group A?
j What do you think their results show?
k What conditions are being tested by group B?
l What do you think their results show?

Things to do

1 Copy and complete:
Water is taken into a plant through its . . . .
It is then carried up the . . . . in tiny tubes called . . . . Water is lost from the plant in the . . . . and more . . . . is drawn up the . . . . from the roots.
Root hairs also absorb . . . . from the water in the soil.

2 Leaves lose water, just like washing on a line.
In what sort of conditions do you think leaves would lose i) most water?
ii) least water?
Try to explain this.

3 Weeds have strong roots that anchor them in the soil.
Design a piece of apparatus to test how strong the roots of some common weeds are.

4 How strong is a stem? Suppose you are given 2 different pieces of stem each 10 cm long. Plan an investigation to find out which is the stronger.
Include a diagram of the apparatus you would use.

It's all in the balance

Learn about:
- the carbon cycle
- deforestation and its effects

The composition of the air that we breathe is pretty constant: about 78% nitrogen, 21% oxygen and 0.04% carbon dioxide.

▶ Copy and complete the word equation for photosynthesis:

carbon dioxide + ... $\xrightarrow{\text{light energy}}$ glucose (sugar) + ...

▶ Copy and complete the word equation for respiration:

. . . . + oxygen ⟶ water + ... + energy

a Which process takes in carbon dioxide from the air?

b Which process puts carbon dioxide back into the air?

c Which process takes oxygen out of the air?

d Which process puts oxygen back into the air?

e What does this tell you about the importance of plants in balancing the composition of the air?

The carbon cycle

▶ Copy and complete the diagram of the carbon cycle, using the words in the box opposite:

animals photosynthesis plants respiration feeding

The carbon cycle

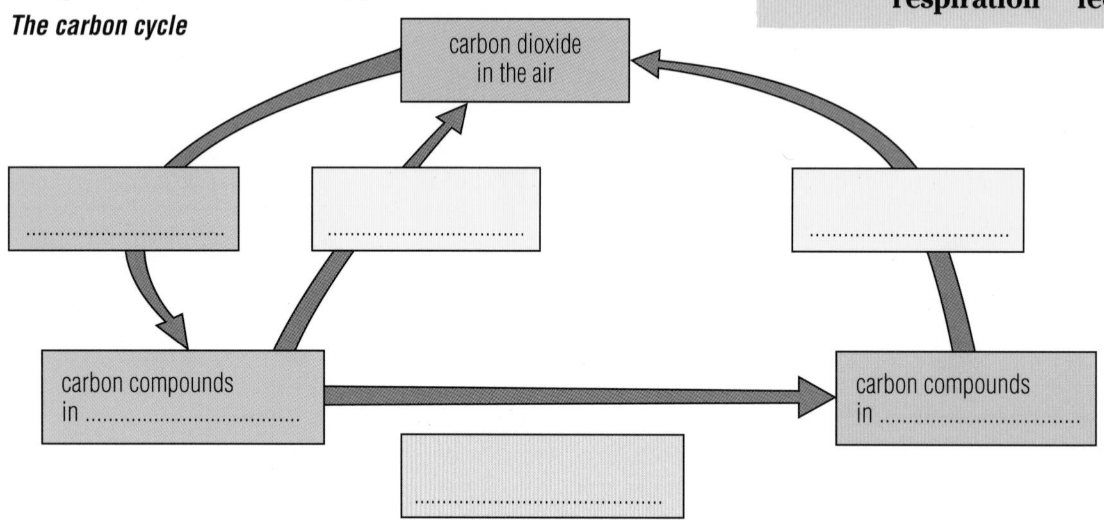

Bubbles in a pond

The graph shows the concentration of 2 gases measured in a pond over 4 days:

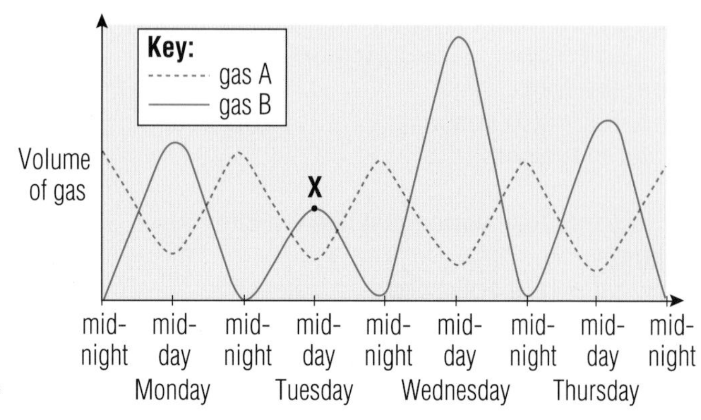

f i) What was gas A?
ii) Explain the shape of the graph for gas A.

g i) What was gas B?
ii) Explain the shape of the graph for gas B.

h Why do you think that there was a low peak at point X?

Amazonia

Brazil has the largest area of rainforest on Earth.
But deforestation is taking place at an alarming rate.
Deforestation reduces the rainforest by 15 000 km^2 a year.
Only 60% of the original forest remains.

Your task is to argue the case either for or against the clearance of the rainforests (even if you disagree with the role that you are given). Use some of the points above and the information on the Help Sheet.

Things to do

1 Copy and complete:
The balance of carbon dioxide and . . . . in the atmosphere is controlled by . . . . and respiration. Carbon dioxide is taken in by . . . . and converted into . . . . compounds during photosynthesis. Plants may be . . . . by animals. Plants and . . . . return carbon dioxide to the atmosphere during . . . .

2 Use the library, an encyclopaedia or the internet to research the following:
a) deforestation b) desertification
c) endangered hardwoods
d) plants as medicines.

3 Matthew decided to make a bottle garden. Inside the bottle he grew some plants and put a butterfly in. He knew that butterflies feed on nectar, so he also put in a dish of sugar solution.

a) Why was the butterfly able to breathe oxygen inside the bottle?
b) Explain what would happen to the butterfly if no plants were put into the bottle.
c) Apart from its droppings, what did the butterfly make that helps the plants to grow?

Questions

1 Label the parts of the leaf using the following words: palisade cells, air spaces, epidermis, stoma, waxy layer, spongy cells, chloroplasts.
Write down: A = chloroplasts, etc.

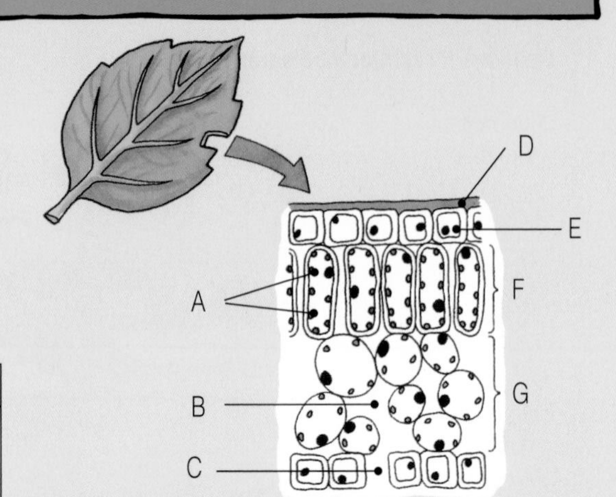

2 Sanjit shone different amounts of light on some pondweed. He recorded the number of bubbles of gas given off by the pondweed per minute:

Units of light	Number of bubbles per minute
1	6
2	14
3	21
4	24
5	26
6	27
7	27

a) Draw a line-graph to display his results.
b) How many bubbles of gas would you expect the plant to make at
i) 2.5 units of light? ii) 8 units of light?

3 The apparatus shown in the diagram was used to measure how much water was lost from the leaves in 24 hours. The apparatus was weighed at the start and at the end of the experiment.
a) Explain how you think the apparatus works.
b) What do you think the results would show?
c) What do you think the oil is for?

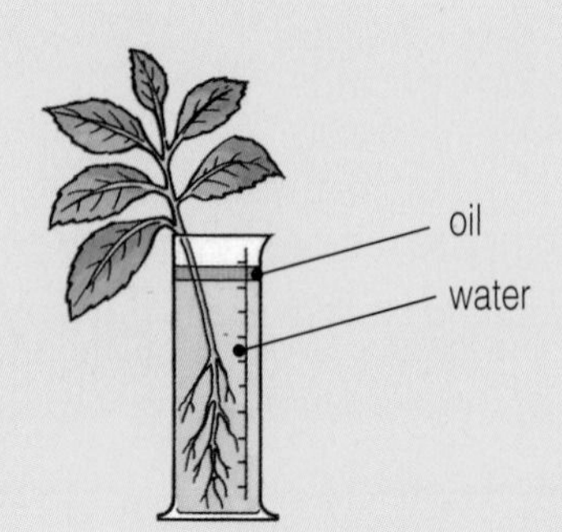

4 Say if you think the following statements about photosynthesis are true or false:
a) Plants get all their food from the soil.
b) Food is transported around a plant in the phloem.
c) Water for photosynthesis is absorbed through the leaves.
d) Chloroplasts are found inside all plant cells.
e) Oxygen is a waste product of photosynthesis.
f) Nitrates are needed for plants to make protein.

5 Potatoes are full of starch.
But the starch is made in the leaves.
So how do you think the starch gets into the potatoes in the soil?
Draw a diagram of a potato plant, with arrows, to show how this happens.

6 Environmental factors can affect the rate (how fast) of photosynthesis.
Which factors slow the rate of photosynthesis of a wheat crop:
a) on a cloudy day?
b) on a sunny day in early March?
c) on a sunny summer's day?
d) if the soil is waterlogged?

Plants for food

Plants are amazing 'energy transformers'. They can change light energy from the Sun into stored chemical energy. Then animals eat the plants and convert the chemical energy into energy that they can use. All the elements in your body can also be traced back to plants.

In this unit you can learn more about how we use plants for food.

Cycles

Learn about:
- decomposers
- nutrient cycle
- sustainable development

Can you remember what happens to dead plants and animals?
They rot away. We say that they **decompose**.
The microbes that make dead things rot are called **decomposers**.
The most important microbes are **fungi** (moulds) and **bacteria**.

▶ Look at the diagram:

a How do plants take up nutrients?

b How do nutrients get into animals?

c How do nutrients get back into the soil?

d Name the producer in the diagram.

e Which is the consumer? Is it a herbivore or a carnivore?

f Think about your last dinner.
Draw a food web that includes a 'human'.

g How could you show that a potato contains starch?

h Where does the starch in the potato come from?

i Name some other parts of plants we eat for starch.

j Why do plants store starch?

The natural roundabout

When plants and animals die, they decompose.
Nutrients are then put back into the soil.
Without fungi and bacteria, the dead material would never decompose.

▶ Look at the diagram:

k Which process makes soil nutrients part of green plants?

l What are the living things that return nutrients to the soil?

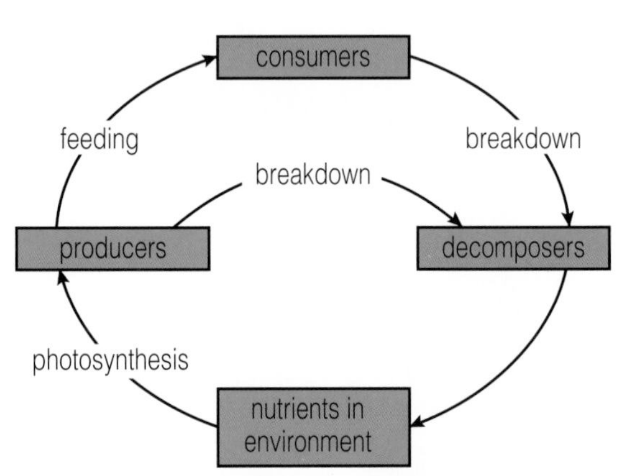

We call the movement of nutrients a **nutrient cycle**.
This takes place on land, in fresh-water and in the sea.

m Where do you think the decomposers are found in the sea?

n Can you name 3 of the most important nutrients?

Natural or chemical?

Chemical fertilisers contain the nutrients **nitrogen**, **phosphorus** and **potassium**. They are easy to store and to use. The farmer also knows exactly how much of each nutrient is being used. However, sometimes chemical fertilisers get washed out of the soil into rivers.
Some farmers prefer to use **natural fertilisers** like manure.
They rot down slower and add **humus** to the soil to improve it.

o Can you give 2 advantages of organic fertilisers?

p Can you give 2 advantages of chemical fertilisers?

Breaking the cycle

Humans share the Earth with millions of other living things.
Unfortunately some human activities destroy wildlife.
They disrupt the natural cycle of life that keeps environments in balance.

- Pollution can harm our environment in many ways.
- Clearing land for crops, housing, roads, factories and mines destroys the habitats of plants and animals.
- Over-fishing can cause the collapse of fish stocks.
- Hunters and poachers are threatening the survival of rare species.

Do some research into ways in which we can protect living things and their environments.
You could find information in books, ROMs or on the internet.

Make a leaflet aimed at convincing others of the need to protect the environment.

Your ideas should:

- be workable and able to be enforced;
- not threaten the livelihood of people, like fishermen;
- use the idea of **sustainable development**, such as replacing trees that have been felled for timber or not taking so many fish out of the oceans that they cannot be replaced by natural reproduction

Things to do

1 Copy and complete:
Decomposers are microbes that . . . . down dead things. The most important decomposers are . . . . and bacteria. The nutrients in the soil are replaced when . . . . and animals die and . . . . Plants take up these nutrients and use them during . . . . Three important nutrients for plant growth are . . . . , phosphorus and . . . .

2 Recycling means using materials again.
a) Make a list of materials that can be recycled.
b) What effect could recycling of these materials have on:
 i) The raw materials used to manufacture goods?
 ii) The energy needed to manufacture these goods 'from scratch'?
 iii) Landfill sites?

3 35% of all our waste comes from packaging.
a) Make a list of the different types of packaging that we take away from shops.
b) In what ways does this packaging cause us problems?

Every year every person in the UK uses up one tree's worth of paper.
c) How would the recycling of paper help our environment?

4 Use books, ROMs or the internet to find out how the following help conservation in the UK:
a) National Nature Reserves (NNRs)
b) Sites of Special Scientific Interest (SSSIs).
c) Nitrate sensitive areas.
d) Heritage Coasts.
e) Set-aside.
f) Tree preservation orders.

Plant growth

Learn about:
- essential nutrients
- fertilisers

A lot of our food comes from plants.
Think about what you have eaten over the last 24 hours.

▶ Write down the food that you think came from plants.

Plants for food

Farmers try to grow enough food for us to eat.
They try to give their crops the best conditions for growth.

a Make a list of things you think plants need to grow well.

▶ Look at the photographs of lettuces growing in a greenhouse: Those in ***A*** are growing in air which has more carbon dioxide than those in ***B***.

b Which do you think would sell for the best price?

c Can you explain why these lettuces are bigger?

d If you were growing a crop in a greenhouse, how could you:
i) increase the amount of time that the plants are in the light?
ii) keep the plants at the right temperature?

Fertilisers

Plants also need chemicals called **nutrients** for healthy growth.
These are usually found in the soil.
They are taken up in small amounts by the roots of the plant.
If the soil does not contain enough nutrients, then the farmer adds more as **fertilisers**.

▶ Look at the table showing the effects of fertilisers on wheat:

Fertiliser	*Nitrogen added*	*Phosphorus added*	*Potassium added*	*Wheat yield (tonnes per hectare)*
none	×	×	×	1.70
A	✓	×	×	3.80
B	×	✓	✓	2.00
C	✓	✓	✓	7.00

e Which fertiliser gives the biggest increase in growth?

f Which nutrients are found in fertiliser B?

g Which nutrient do you think is the most important for wheat growth?

NPK fertilisers contain:

- Nitrogen (N), it is used to make proteins for growth,
- Phosphorus (P), for healthy roots,
- Potassium (K), for healthy leaves.

The proportions of nitrogen, phosphorus and potassium (N : P : K) are shown on the fertiliser bag.

h Which nutrient is missing from the fertiliser in the picture?

i Another fertiliser is called 25 : 5 : 5. What do you think this means?

NPK fertilisers

Give more, grow more

We can use liquid fertiliser (plant food) to grow good house plants.

Plan an investigation to see how the growth of some duckweed depends on the fertiliser.

You could look at the effects of either
- different fertilisers

or
- different strengths of the same fertiliser.

- What are you going to measure to show growth?
- What things will you need to keep the same if it is to be a fair test?
- What factors can't you control? How will you deal with these?
- What are you going to change?
- How long will your investigation last?
- How often will you take measurements?

Show your plan to your teacher before you try it out.

- After carrying out your enquiry, draw conclusions from your results. What confidence do you have in your conclusions?
- How could you change the method you used to improve the quality of data collected?

Where there's muck, there's growth

Why do you think gardeners put horse manure on their roses?

Animal waste is broken down by microbes.
Some bacteria get nitrogen out of the waste.
Plants take up the nitrogen through their roots and use it for growth.

j How else do you think manure ***improves*** the soil?

k Why do you think some gardeners prefer to use chemical fertilisers?

▶ Find out what you can about **organic gardening**. You could find information in books, ROMs or on the internet.

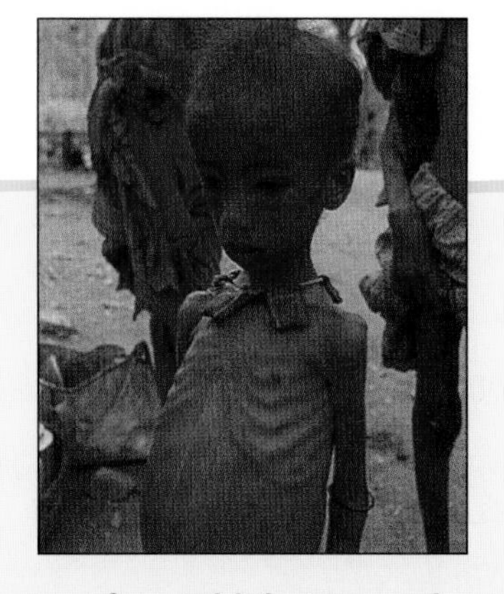

Things to do

1 Copy and complete:
To grow, plants need . . . . from the air, water from the . . . . , and sunlight. They also need nutrients from the soil. These include N (. . . .), P (. . . .) and K (. . . .). If the . . . . does not contain enough nutrients, the farmer adds . . . .

2 Animal manure and compost are both good for plant growth.
Explain why you think this is true.

3 What nutrients are found in NPK fertilisers?
If a fertiliser has an NPK value of 10 : 5 : 10, what does it mean?

4 We are often told that enough grain is grown to feed everyone in the world. So why do you think it is that people are starving?
Use the following clues to explain why there is hunger in the world:
a) transport b) wars c) food mountains d) pests e) drought.

Dicing with death

Learn about:
- toxins in food chains
- pesticides

Pesticides are chemicals that farmers use to kill **pests**.

a What kinds of plants and animals might be pests to a farmer?

b Why do you think farmers need to kill pests?

The main farm pests are insects, weeds and moulds. Farmers need to control them because they destroy crops.

Nasty stuff

DDT is a pesticide. Only a small amount is needed to kill ***any*** insect. It was used to kill plant pests and the mosquitoes that spread **malaria**. However, DDT does not break down quickly. It stays in the soil for a long time. So it can be passed along food chains. This is dangerous.

This lake in California was sprayed with DDT to control midges.
Look at the diagram to see what happened:

c How does DDT get into this food chain?

d What happens to the concentration of DDT as it passes along the food chain? Try to explain this.

e Why are the fish-eating birds the first to be killed by DDT?

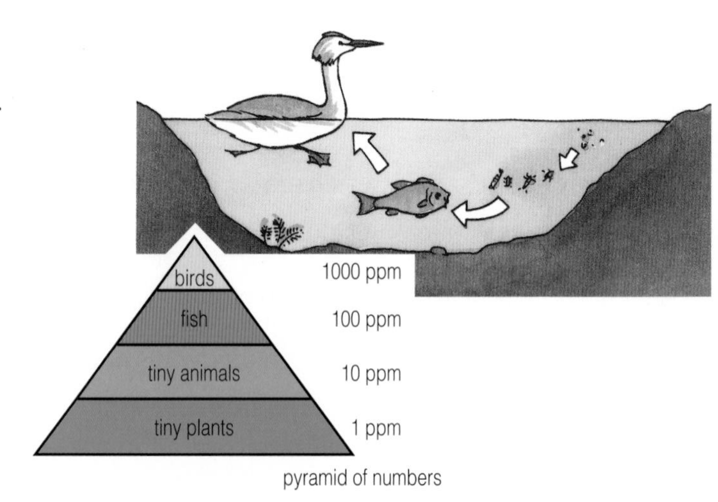

pyramid of numbers

The perfect pesticide?

A pesticide must be effective, but it must also be safe. Discuss, in your groups, what makes a good pesticide.

- Which insects should it kill?
- For how long should it be active?
- Should it dissolve in water? Why?
- How will it get into the insect's body?
- On which part of the body will it act?
- How will it be safely used?

Write down your ideas.

Carry out some research and produce a table showing advantages and disadvantages of using pesticides.

- Were the authors of the information you found biased? Why?

Weed survey

Imagine that your school field is a farmer's field.
You have to identify the weeds found on the field.
To do this, use a Help Sheet, or the pictures shown here.
Then write a report for the farmer which will outline:

- the population of each weed in the field,
- what problems these weeds will give the farmer,
- how to treat the field to get rid of the weeds,
- possible consequences of removing the weeds using the method that you suggest.

plantain

daisy

Who killed the sparrowhawk?

In the 1960s, seeds were often dipped in pesticide to protect them from pests.
Soon birds of prey, like the sparrowhawk, started dying.
Their bodies had large amounts of pesticide in them.
They also laid eggs with thin shells.

f How do you think the pesticide got from the seeds into the sparrowhawk**?**

g Why was it less concentrated in the bodies of seed-eating birds**?**

h Why do more birds die if their eggs have thin shells**?**

Pesticides, like DDT, are now banned in many countries.

Things to do

1 Copy and complete:
Chemicals that kill pests are called . . . . Pests that can damage crops are . . . . , moulds and . . . . Some pesticides do not . . . . down easily. They can enter food . . . . and, as they are passed on, they become . . . . concentrated. Animals at the . . . . of the food chain are the first ones to die.

2 A new pesticide has been made to kill weeds.
The manufacturer wants to know at what concentration to sell it.
If it is too strong, it will kill the crop and harm wildlife.
If it is too weak, then it will not kill the weeds.
Plan an investigation to find out the best concentration of pesticide.

3 Some insects can develop **resistance** to a particular pesticide.
a) What is likely to happen to the farmer's crop if this happens?
b) What could the farmer do about it?

4 In Holland, they are using ladybirds to kill off lice on trees. The ladybirds are imported from California. Sixty million are being sold to Holland this year. The US suppliers say "Unlike pesticides, they are environmentally-friendly and real cute too!".
a) This is an example of **biological control**. What do you think this means?
b) What are the advantages of using ladybirds instead of pesticides?

Against all odds

Learn about:
- sources of pollution
- effects of pollution
- growing plants

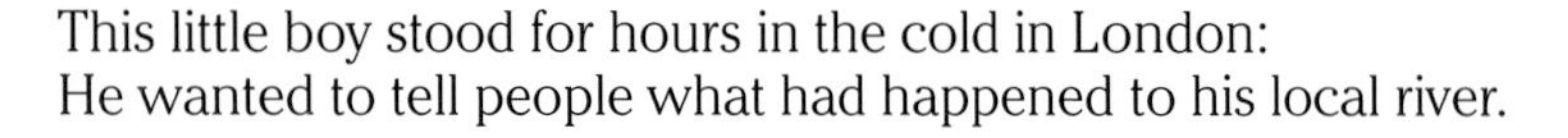

This little boy stood for hours in the cold in London:
He wanted to tell people what had happened to his local river.

What do you think could have happened to the river?

This is just one example of **pollution**.
What do you think we mean by pollution?
Write down some of your ideas.

Pollution is when we do things that harm our environment.

What's causing the pollution?

Look at these photographs:

▶ Copy the list below of the types of pollution shown in the photographs.
Match each one with the correct effects from the other list:

Type of pollution	*Effects of pollution*
• dumping radio-active waste	• damages body cells
• oil spills	• lung diseases
• dangerous tips	• kills trees and water life
• smog	• kills sea-birds
• detergents	• harmful chemicals leak out
• acid gases	• too many water plants grow

Signs of pollution

When fossil fuels burn, they make acid gases like **sulphur dioxide**.
These dissolve in water in the clouds to make acid rain.
Acid rain can kill plants and fish.

Black spot is a mould that grows on roses. The mould cannot live if there is sulphur dioxide in the air.

a What does it tell you if roses do not have black spot?

Lichens are plants that are sensitive to sulphur dioxide in the air.

b Look at the shrubby lichens in the picture.
What do you think the air is like?

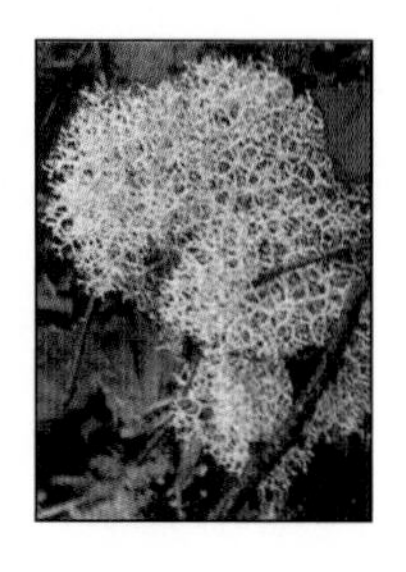

Perfect conditions

Have you ever visited the Eden Project in Cornwall?
At the site you will find a variety of climates recreated in huge domes called 'biomes'. Inside these giant conservatories, scientists carefully control temperature, humidity and soil types to grow plants from all over the world.
In the biggest biome (which is 240 m long, 110 m wide and 50 m high), the conditions are set like a tropical rainforest. It contains over 1000 species of plants. The biomes are made from special glass that lets ultra-violet light from the Sun pass through it.

Design a glasshouse

Commercial growers in many countries have set up huge glasshouses to grow plants more effectively.

- Suggest some advantages and disadvantages of growing crops in glasshouses.
- Design your own large-scale glasshouses for growing crops. Draw your design, labelling its special features and explaining how you would control the environment inside.

Things to do

1 Copy and complete:
Pollution occurs when . . . . put harmful . . . . or energy into the . . . .
When fossil fuels are burned, they give off gases such as . . . .
Detergents and . . . . can drain into rivers and cause too many . . . . to grow.
Sea-birds can be killed as a result of . . . .
. . . .

2 a) Name 3 conditions that can affect the rate of photosynthesis.
b) How could you control each of these 3 conditions in a glasshouse:
i) in summer? ii) in winter?

3 Look at the graph:
It shows how the rate of photosynthesis can be changed by environmental conditions.

a) What effect does increasing the light intensity have on the rate of photosynthesis?
b) What environmental factor is limiting the rate of photosynthesis at point X on the graph?
c) What could you do to increase the rate of photosynthesis at point X?

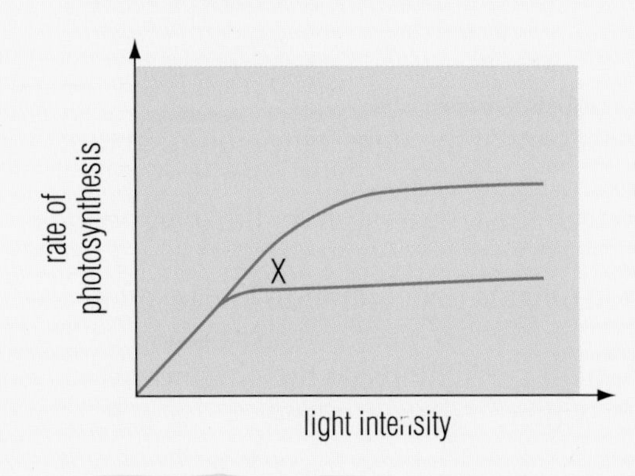

Questions

1 Which parts of the following plants do we eat:
a) pea b) wheat c) lettuce d) coconut
e) banana f) rice g) radish h) onion
i) carrot j) cabbage k) potato l) tomato?

2 Draw a concept map, labelling the connections between the following terms:

producer, consumer, food web, photosynthesis, energy, Sun, glucose, starch, root, leaf, stem

3 The table shows the effects of adding nitrate fertiliser on the growth of a wheat crop.

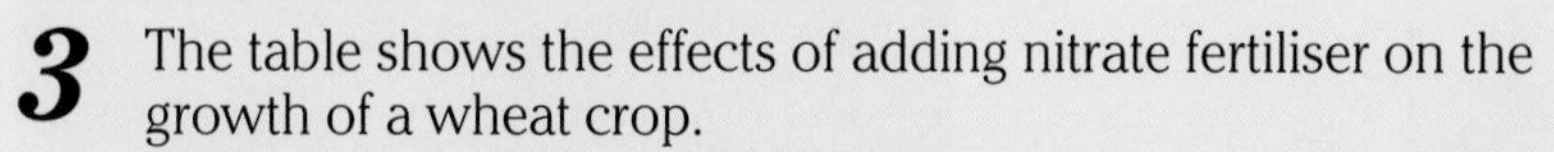

Nitrate fertiliser added (kg/ha)	0	50	100	150	200	250
Yield of wheat (tonnes/ha)	2	6	7.5	8.5	9	9

a) Draw a line-graph to display the results.
b) How much fertiliser would you use to get the best wheat crop?
c) Give 2 reasons why you should not use ***more*** fertiliser than this.

4 Farmer Jenkins put lots of fertiliser on his fields in the autumn.
The following summer the river nearby was full of water weeds.
Some of the weeds died and started to rot.
This took oxygen out of the river.
a) Why do you think the water weeds grew so much?
b) What do you think happened to the fish in the river?
c) How would you solve this problem?

5 a) Why are pesticides used?
b) What are the alternatives to using traditional chemical pesticides?
c) Should pesticides be used to produce more food at the expense of other animals?
Explain your answer, presenting both sides of the argument.

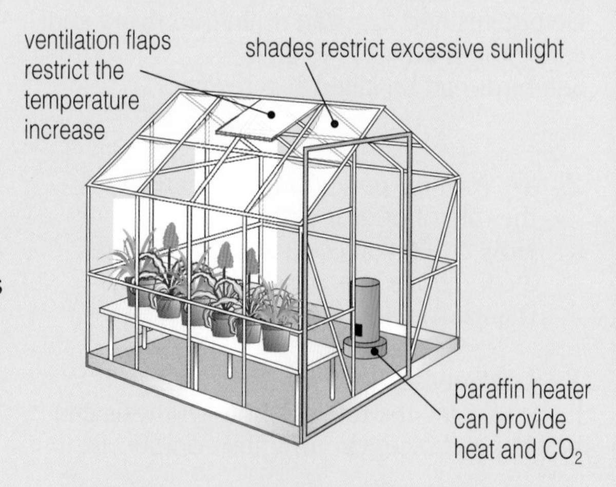

6 Look at the diagram of the glasshouse:
a) How are the plants kept at the right temperature in the winter?
b) Give ***2*** ways in which the temperature in the glasshouse is controlled in summer.
c) How is the carbon dioxide concentration increased inside the glasshouse?
d) How is the amount of sunlight inside the glasshouse controlled i) in winter? ii) in summer?

Metals and metal compounds

Just imagine life without metals – no iron for steel used in making cars, trains, bikes; no copper for electrical wiring and water pipes.

In this unit you can find out more about the properties of metals and their reactions.

Metals in use

Learn about:
- properties of metals
- uses of metals

Most elements can be sorted into 2 sets – **metals** and **non-metals**.

a What is an element?

b In the periodic table, where are the metals?

c Make a list of the names of some common metals.
Write the symbol for each metal next to its name.

d Make a list of names and symbols of common non-metals.

▶ How much do you remember about the properties of metals?
Write down your ideas. You could make a list or mind map.

Looking at the properties of metals

Appearance

Rub each metal with a piece of sandpaper to clean it.
What does the clean metal look like?

e How do metals get tarnished?

Melting point

Heat an iron nail strongly in a Bunsen burner flame.
Does it melt?

f What is the melting point of iron?
(Use a data book to find this.)

Do metals conduct electricity?

Set up the circuit opposite:
If the bulb lights, the metal must conduct.

g Do all the metals conduct electricity?

h Can you tell which is the best conductor of electricity?

battery
bulb
element to be tested

Do metals conduct heat?

Set up the apparatus opposite:
When the heat reaches the end of the rod, the grease melts.
When the grease melts the pin falls.
Test the metals. Which pin falls first?

i Which metal is the best conductor of heat?

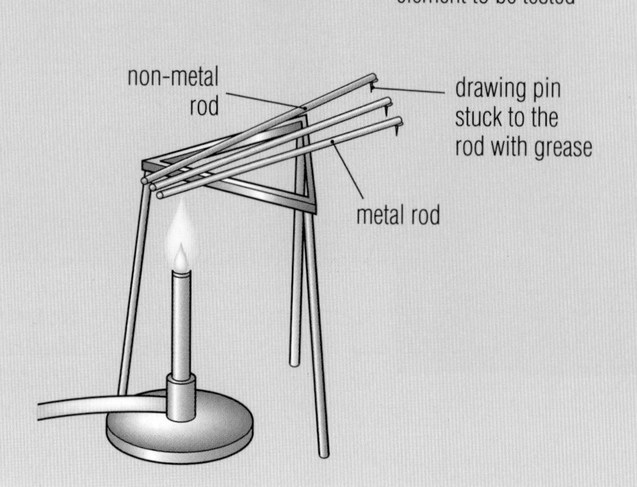

When you have tested your metals, see your teacher.
You may be able to compare non-metals in your conductivity tests.

Comparing metals and non-metals

▶ Make a summary table to show the differences between the properties of metals and non-metals:

Property	Typical metal	Typical non-metal

These are the usual properties. Be careful! There are some exceptions which your teacher may show you.

Lithium, sodium and potassium are metals. They are soft. They have low melting points.

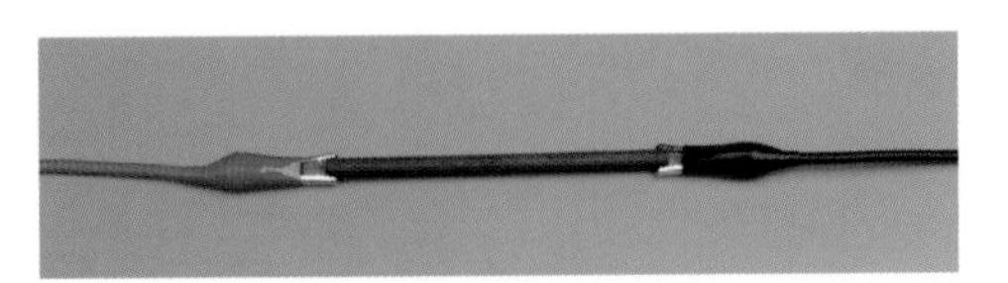

Graphite is a form of carbon (a non-metal). It conducts electricity.

Diamond is a form of carbon (a non-metal). It has a very high melting point.

j Why do you think metals are so useful? (Which properties are very helpful?)

Setting up a database

Make a class database about metals.
Choose one metal to research.
Try to find answers to the following questions:

- What is its chemical symbol?
- Where do we get the metal from?
- What is it used for?
- What is its density?
- What are its melting point and boiling point?
- How well does it conduct heat?
- How well does it conduct electricity?
- How much does it cost?

Things to do

1 Copy and complete, using words from the box. The words could be used once, more than once, or not at all.

good high low strong
shiny dull brittle

Metals are . . . . conductors of heat and electricity. They usually have . . . . melting points and boiling points. They are . . . . when clean so they reflect light. They are . . . . so can be used to make bridges.

2 The following words can also be used to describe metals.
Find out what each word means.
a) Sonorous b) Malleable c) Ductile

3 A few elements like silicon are called ***metalloids*** or ***semi-metals***.
What does this mean?
Find the names of some other semi-metals.

4 Say whether the following statements are true (T) or false (F):
a) All metals are solids.
b) All non-metals are gases.
c) All metals have high melting points.
d) All non-metals are poor conductors of electricity.
e) All metals conduct electricity.
f) All metals are magnetic.

Making salts

Learn about:
- some metals reacting with acid
- making crystals of a salt

The word ***salt*** probably makes you think of something you put on your chips!
But in science a **salt** is much more than this.

In this lesson you can make lots of ***different*** salts.
To do this you will start with an **acid** and a **metal**.

a Write down everything you know about acids.
Include these words if you can.

pH   alkali   strong   red   weak   neutral

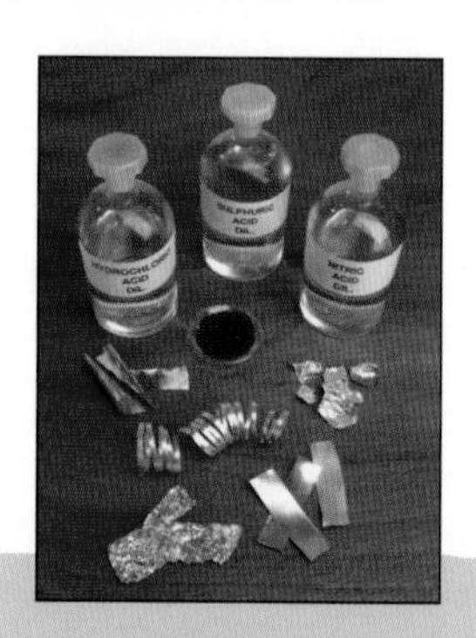

b Write down everything you know about metals.
Include these words if you can.

shiny   conductor   strong   electricity   heat   hard

Reacting acids with metals

Try these tests with hydrochloric acid and metals.
Make a table to record your results.

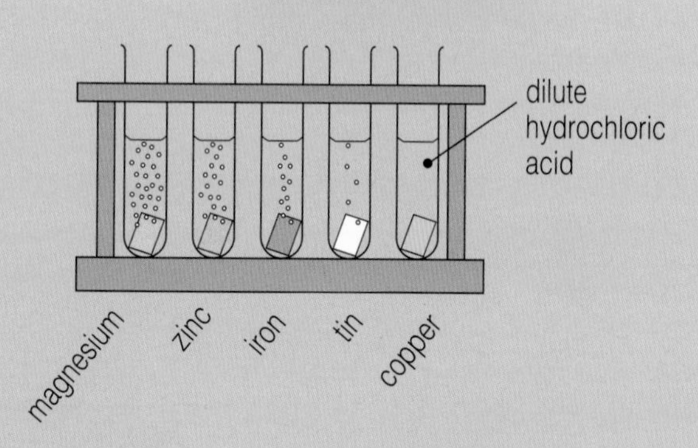

Put 5 test-tubes in a rack.
Put about 2 cm^3 of dilute hydrochloric acid in each test-tube.
Clean the metal samples with sand paper.
Add the metal samples to the acid as shown.

c How do you know if there is a reaction?
d What is the name of the gas given off?
e How can you test for this gas?

If the metal does not react, put the tube in a beaker of hot water.
See if the metal reacts with ***warm*** acid.

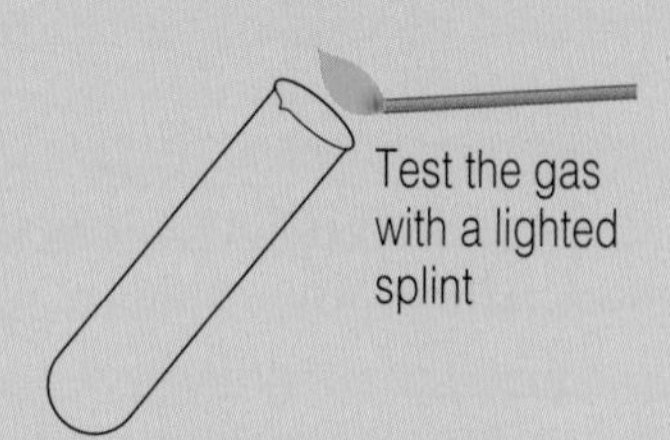

f You could use dilute sulphuric acid now instead of dilute hydrochloric acid.
What do you think would happen?
Ask your teacher if you can check your idea.

g What do you think would happen with dilute nitric acid and metals?

When a metal reacts with an acid a **salt** is made.

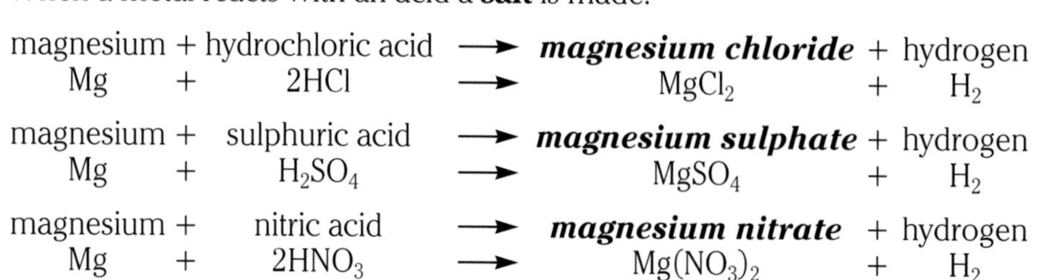

magnesium + hydrochloric acid ⟶ ***magnesium chloride*** + hydrogen
$Mg + 2HCl \longrightarrow MgCl_2 + H_2$

magnesium + sulphuric acid ⟶ ***magnesium sulphate*** + hydrogen
$Mg + H_2SO_4 \longrightarrow MgSO_4 + H_2$

magnesium + nitric acid ⟶ ***magnesium nitrate*** + hydrogen
$Mg + 2HNO_3 \longrightarrow Mg(NO_3)_2 + H_2$

sulphuric acid is H_2SO_4
hydrochloric acid is HCl
nitric acid is HNO_3

When some or all of the hydrogen in an acid is replaced by a metal, we get a compound called **a salt**

Salts made from **sulphuric acid** are **sulphates**.
Salts made from **hydrochloric acid** are **chlorides**.
Salts made from **nitric acid** are **nitrates**.

Making zinc sulphate

An acid can be changed to make a salt. The change is a ***chemical reaction.***

⚠ acid – eye protection

Carry out this experiment to make crystals of a salt. You can use the Help Sheet.

Add zinc powder to dilute sulphuric acid and stir.

→

When no more zinc dissolves, filter the mixture.

→

Carefully evaporate the filtrate to half its original volume.
Then leave it to cool.

⚠ eye protection

sulphuric acid **(acid)**	+	zinc **(metal)**	⟶	zinc sulphate **(a salt)**	+	hydrogen

h Look at the photos. Which one shows the hydrogen being made?

i Why should you keep adding zinc until ***no more dissolves***?

j What do you see when you evaporate and then cool the solution?

k What is the name of the solid made? This is the product.

l Try to write a symbol equation for the reaction.

Acids are corrosive

Things to do

1 Copy and complete:
a) Acids have pH number . . . . than 7.
b) Acids react with metals to make a . . . . and hydrogen.
c) acid + metal ⟶ a salt + ...
d) Hydrogen gas . . . . with a . . . . spill.
e) Sulphates are made from . . . . acid.
f) Chlorides are made from . . . . acid.
g) Nitrates are made from . . . . acid.
h) To make zinc chloride we use . . . . and . . . . acid.

2 Acid and metal reactions are used to make lots of salts. Why can't we make copper sulphate this way?

3 Complete these word equations:
a) magnesium + nitric acid ⟶
b) iron + sulphuric acid ⟶

4 Some solutions were tested with pH paper:

solution	A	B	C	D	E
pH value	9	1	5	13	7

1 2 3 4 5 6 7 8 9 10 11 12 13 14

a) Say whether each solution is acidic, alkaline or neutral.
b) What colour does the pH paper turn with i) D? ii) E? iii) B?
c) Which solution is the most acidic?
d) Which solution could be pure water?

5 Acid rain causes problems. Think about the reactions of acids.
a) What happens when acid rain falls on metal?
b) How can we protect metal structures against attack by acid rain?

Acids and bases

Learn about:
- metal oxides plus acid
- metal carbonates plus acid

We can make a **salt** by reacting acid with some metals.

▶ Copy and complete these word equations:

a zinc + nitric acid ⟶

b iron + hydrochloric acid ⟶

c magnesium + sulphuric acid ⟶

Take care with the next one!

d copper + hydrochloric acid ⟶

There are other ways to make salts.
We can react acids with **bases**.
Bases are the oxides, hydroxides and carbonates of metals.

▶ Name the bases with these formulae:

e MgO **f** $CuCO_3$ **g** $Ca(OH)_2$

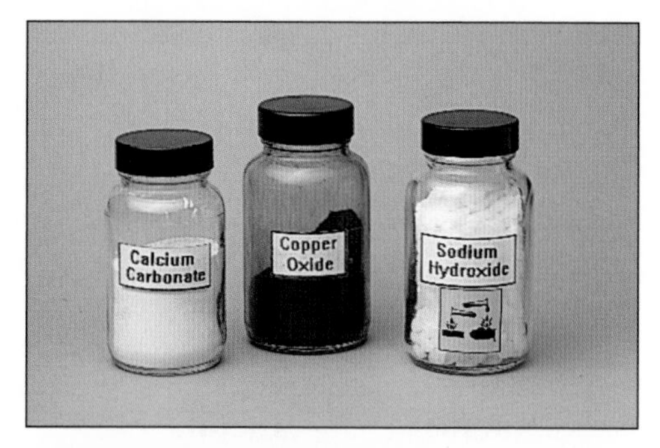

The base **neutralises** the acid.

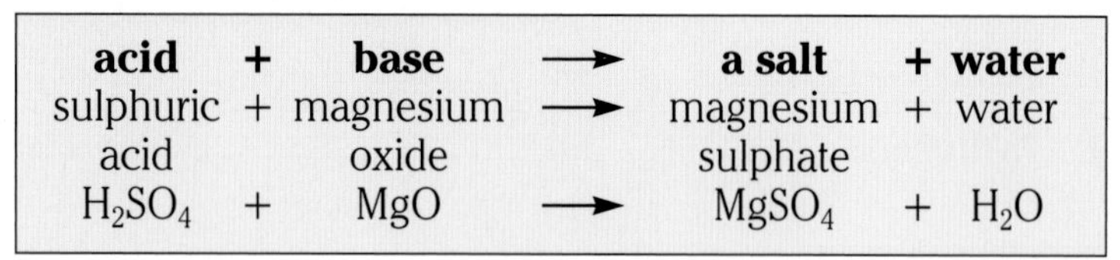

acid	**+**	**base**	⟶	**a salt**	**+ water**
sulphuric acid	+	magnesium oxide	⟶	magnesium sulphate	+ water
H_2SO_4	+	MgO	⟶	$MgSO_4$	+ H_2O

If the base is the metal ***carbonate***, we get another product.

Look at the reaction of magnesium carbonate with acid:

h How can you tell that a chemical reaction is taking place?

i What causes the fizzing in the reaction?

j How can you test for this gas?

sulphuric acid	+	magnesium carbonate	⟶	magnesium sulphate	+	carbon dioxide	+	water
H_2SO_4	+	$MgCO_3$	⟶	$MgSO_4$	+	CO_2	+	H_2O

k Which elements are found in all carbonates?

Your teacher may let you try some other 'carbonate + acid' reactions.
We can use this reaction to prepare carbon dioxide in the laboratory.

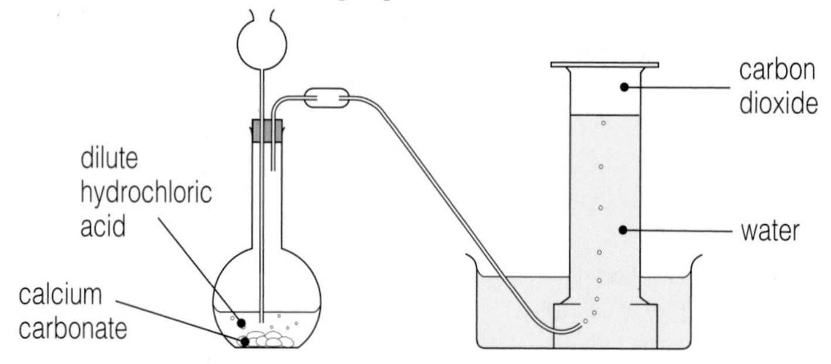

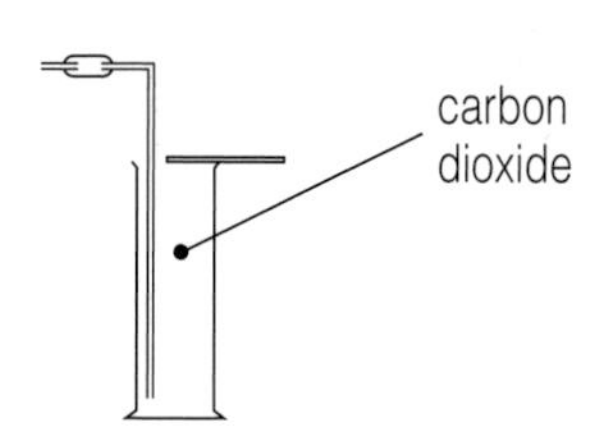

The carbon dioxide can be collected ***over water***. It is only slightly soluble in water.
We can collect it by ***downward delivery***. The gas is more dense than air.
It is colourless, so it is difficult to tell when the gas jar is full.

These are the general word equations for acid + base reactions:

acid + metal oxide ⟶ a salt + water
acid + metal hydroxide ⟶ a salt + water
acid + metal carbonate ⟶ a salt + carbon dioxide + water

Try using an acid + base reaction to make a salt. You can use the Help Sheet.

Making copper sulphate

⚠ acid – eye protection

Add copper oxide to warm dilute sulphuric acid and stir.

When no more copper oxide dissolves, filter the mixture.

Carefully evaporate the filtrate to half its original volume. Then leave it to cool.

⚠ eye protection

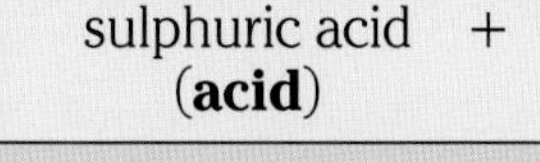

sulphuric acid (**acid**)	+	copper oxide (**base**)	⟶	copper sulphate (**salt**)	+	water

l How can you tell there is a chemical reaction here**?**
m Why should you keep adding copper oxide until ***no more dissolves*?**
n Which substance is left in the filter paper**?**
o What is the name of the solid made**?** This is the product.

If the base is copper carbonate you can still make copper sulphate the same way. (You don't have to warm the acid though.)

sulphuric acid + copper carbonate ⟶ copper sulphate + carbon dioxide + water

p What causes the fizzing in this reaction**?**

Things to do

1 Copy and complete:
a) Bases are the oxides, ... or . . . . of metals.
b) Bases neutralise . . . .
c) acid + carbonate makes . . . . gas
d) . . . . + metal oxide makes a salt.

2 You have 5 test-tubes containing different gases. The gases are:

nitrogen oxygen carbon dioxide
hydrogen air

How can you identify each gas?
Describe the tests and the results.

3 There are 3 bottles and 3 labels! The labels have come off the bottles! You have samples of lime water, copper carbonate and red litmus paper. How can you use these to identify the liquid in each bottle? Describe what you would do.

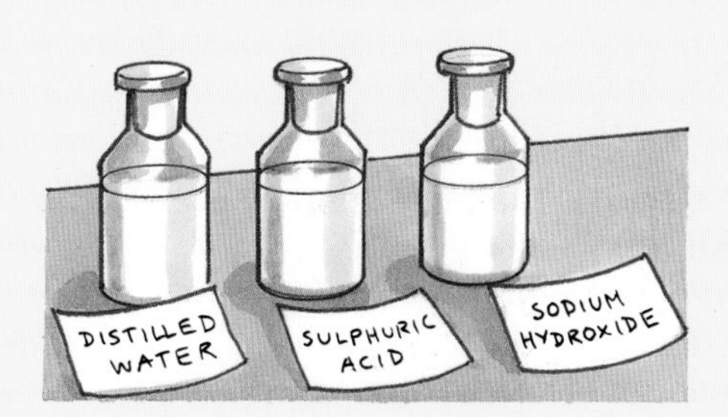

Salt of the Earth

Learn about:
- some uses of salts
- neutralisation using alkali
- making a fertiliser

▶ Make a list of all the salts you have come across in this unit.

a What do all the names of the salts have in common?

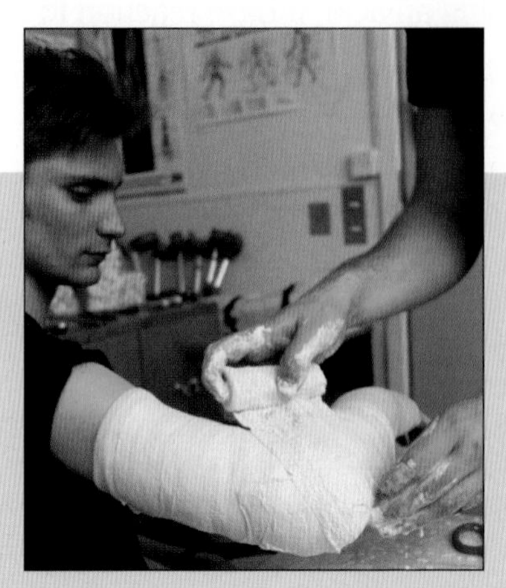

Calcium sulphate is used to set broken bones

Uses of salts

Choose one of the salts from your list or from those given below to find out its uses.

Here are some other examples you might research:

- sodium stearate
- calcium phosphate
- iron sulphate
- silver nitrate
- silver bromide

Some salts can be used as fertilisers.
The 3 main chemical elements in fertilisers are:

nitrogen (N), phosphorus (P) and potassium (K). (See 9D2.)

The elements are in **compounds** in the fertiliser.
On fertiliser bags you often see numbers. These are **NPK** ratios.
They tell you how much nitrogen, phosphorus and potassium the fertiliser contains.
16.8.24 means 16%N, 8%P, 24%K.

▶ Look at these bags of fertilisers:

b Which of these fertilisers:
  i) contain nitrogen?
  ii) contain phosphorus?
  iii) contain all 3 elements?

c What do fertilisers do? Why do we use them?

GROLOTS 21·8·11

Fastgro 14·0·11

Strongro 21·8·0

Making a fertiliser

You can make a simple fertiliser called ammonium sulphate in the lab.

Ammonium sulphate has the formula $(NH_4)_2SO_4$

d Which chemical elements does it contain?

Sodium hydroxide is a base. It is also an alkali.

Copper oxide is a base. It is **not** an alkali.

Ammonium sulphate is a salt. It is made by neutralising an acid.
In Year 7 you learnt that **alkalis** neutralise acids.
An **alkali** is a special kind of base – it dissolves in water.

bases

alkalis

The reaction to make a fertiliser is a **neutralisation**.

acid + alkali (base) ⟶ a salt + water

⚠ Alkaline solutions can be hazardous. ⚠
A concentrated alkali will break down oils and fats in a chemical reaction – and that includes those in your skin and eyes!

Making ammonium sulphate

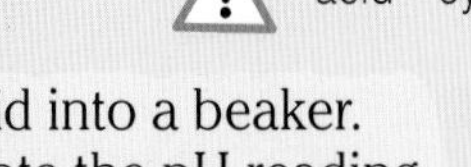

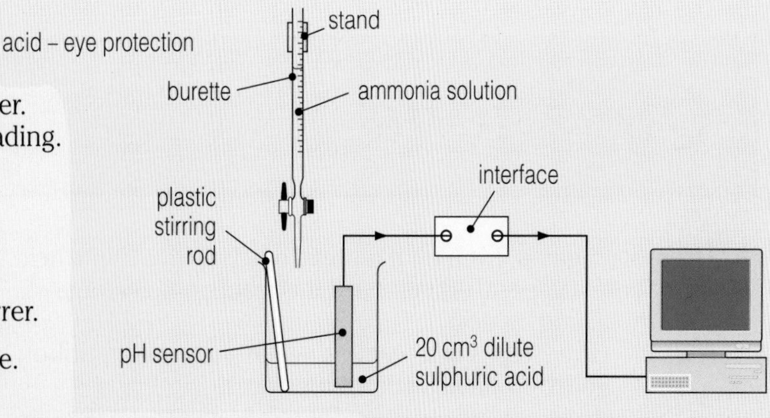

1. Put 20 cm^3 of dilute sulphuric acid into a beaker. Put a pH sensor into the acid. Note the pH reading. Start recording.
2. Fill a burette with ammonia solution. Add 2 cm^3 of ammonia solution to the acid. Stir very carefully with a plastic stirrer. Be careful not to touch the sensor with the stirrer.
3. Keep adding ammonia solution 2 cm^3 at a time. The sensor will measure the pH as it changes.
4. Keep adding ammonia solution until the pH reaches 7 (neutral) or just over. Stop recording. Remove the sensor. The acid has been neutralised. You have a solution of ammonium sulphate.

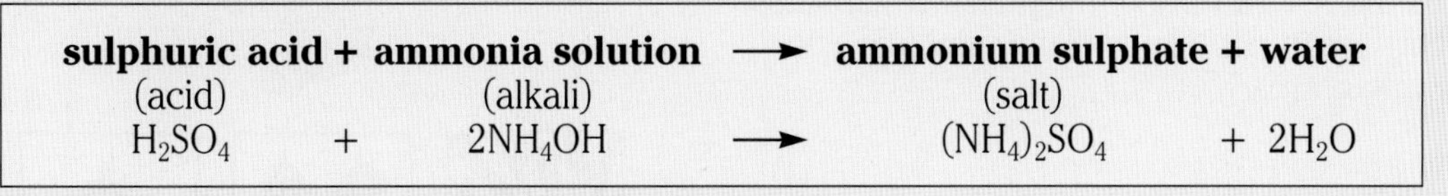

sulphuric acid + ammonia solution ⟶ ammonium sulphate + water

(acid) (alkali) (salt)

$$H_2SO_4 + 2NH_4OH \longrightarrow (NH_4)_2SO_4 + 2H_2O$$

5.

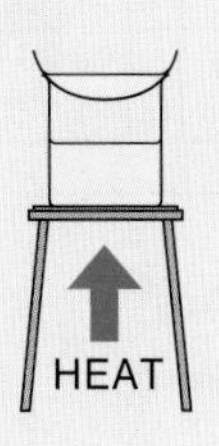

hot

Pour the solution into an evaporating basin.
Carefully evaporate the water from the solution by heating it on a water bath.
Evaporate until only half of the original solution is left.

6.

Leave the basin to cool.
Crystals of ammonium sulphate will form slowly.
Filter the crystals from any solution left.
Dry them between filter papers.

What does your fertiliser look like**?**
Your teacher might give you a sheet to plan an experiment to make your own salt.

Things to do

1 Copy and complete:

a) 3 common elements in fertilisers are . . . . , ... and . . . .
b) NP fertilisers contain . . . . and . . . .
c) An . . . . is a base that dissolves in water.
d) Ammonium sulphate is made from dilute . . . . acid.

2 Look at these NPK values for 4 fertilisers:

0.24.24	15.15.21	25.0.16	27.5.5
1	2	3	4

a) Which fertiliser contains most nitrogen?
b) Which fertiliser contains least nitrogen?
c) Which one is NK fertiliser?
d) Which fertiliser contains the same % of nitrogen as phosphorus?
e) Fertilisers can cause water pollution. Explain why.
f) Plants also need small amounts of:
Ca Mg Na Cu Zn S Fe
Name these elements.

3 Visit a garden centre. Look at packets of fertilisers recommended for growing

i) fruit (e.g. tomatoes)
ii) grass
iii) flowers

a) Make a list of NPK values for these fertilisers.
b) Can you draw any conclusions about the elements needed to grow certain crops?

4 What are ***organic vegetables***? Why do some people want to buy these even if they are more expensive?

Questions

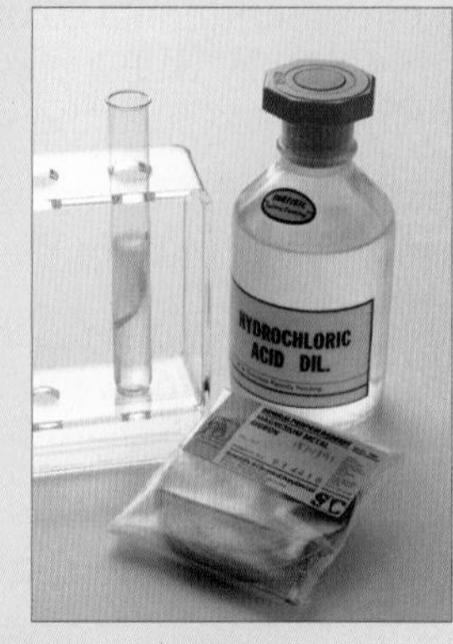

1 Hydrogen gas is given off if a metal reacts with a dilute acid.
a) Write a general equation for the reaction between a metal and an acid.
b) i) Give 2 signs of a reaction when magnesium ribbon is added to dilute hydrochloric acid.
ii) Write a word equation for this reaction.
iii) Write a symbol equation for this reaction.

2 Explain how you could make a sample of magnesium sulphate crystals starting with magnesium ribbon and dilute sulphuric acid. Write a method. Draw diagrams.

3 You have used lots of chemicals to carry out reactions.
Some chemicals are dangerous. They must be used very carefully.
They have hazard labels on their containers.
Look at the hazard labels opposite. Draw sketches of them.
Explain what each hazard label tells us.

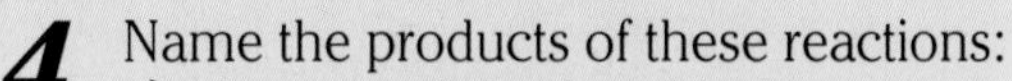

4 Name the products of these reactions:
a) iron + sulphuric acid
b) zinc carbonate + sulphuric acid
c) magnesium oxide + nitric acid
d) sodium hydroxide + hydrochloric acid

5 Stomach powders help to neutralise acids in your stomach.
Plan an investigation to find out which stomach powder is the best value for money.

6 Potassium hydroxide is an alkali.
a) Why is a concentrated solution of potassium hydroxide hazardous to use in experiments?
b) Copper(II) hydroxide is described as a base, not an alkali. Why?
c) Describe the method you could use to prepare crystals of potassium chloride, using potassium hydroxide solution as one of the reactants. What could you do to get your crystals as large as possible?

7 Write 5 multiple-choice questions on the work in this unit.
After each question, show which is the correct answer.

Patterns of reactivity

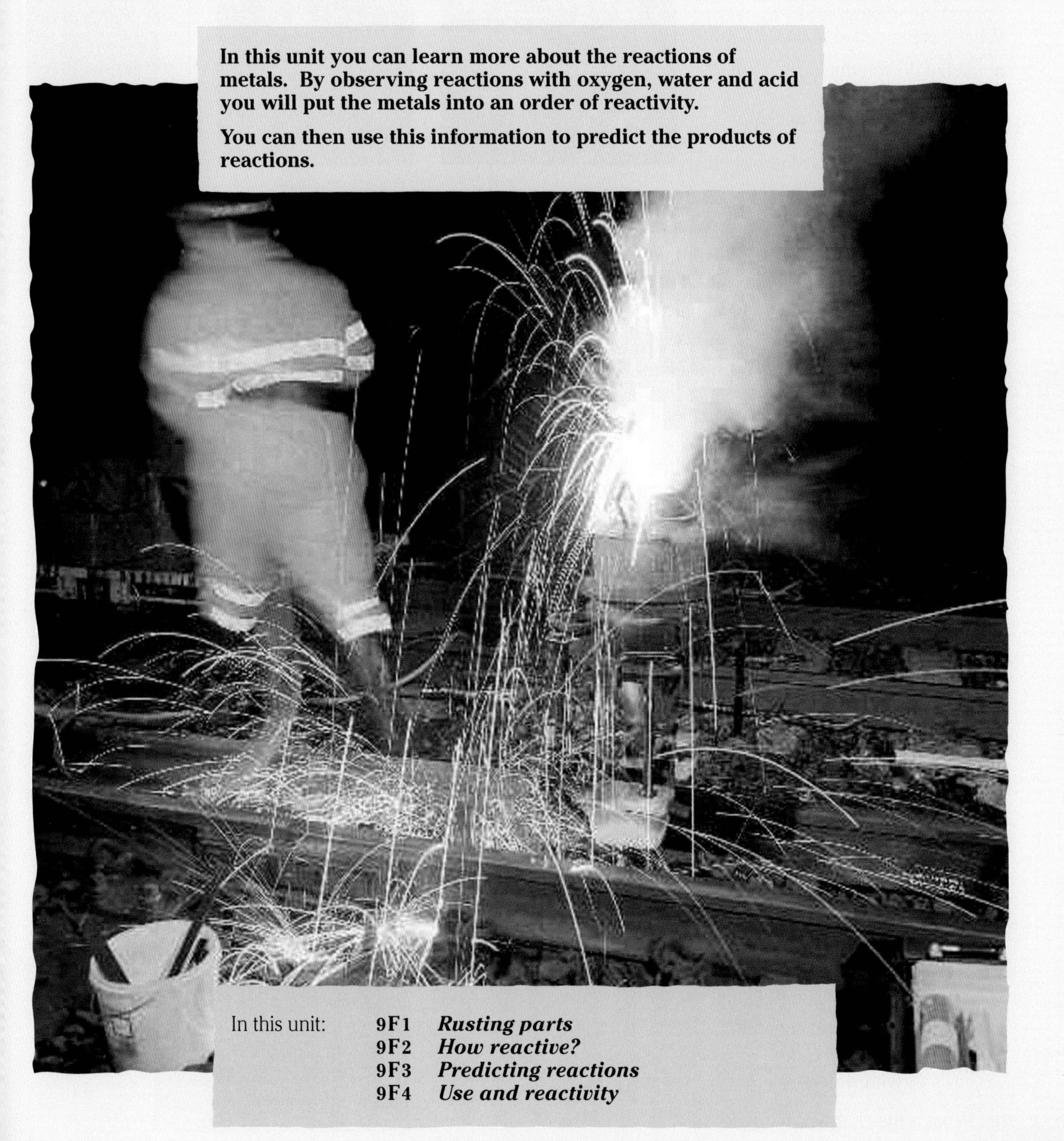

In this unit you can learn more about the reactions of metals. By observing reactions with oxygen, water and acid you will put the metals into an order of reactivity.

You can then use this information to predict the products of reactions.

Rusting parts

Learn about:
- metals tarnishing
- iron rusting
- preventing rust

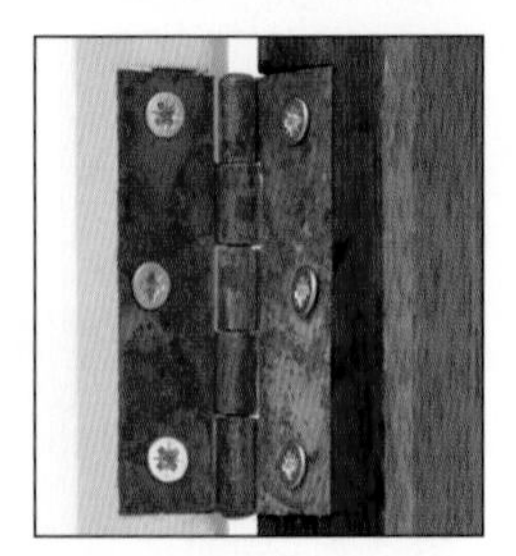

Do metals last forever**?**
Look at the photographs:
What has happened to the metals**?**

a What do you think causes this**?**

Some metals ***tarnish*** quickly.
Others change very slowly.
Do you have examples of tarnished metals at home or at school**?**

Do you have a bike**?** Does anyone in your family have a car**?**
If so, you probably know about the problem of rust!
Every year rust causes millions of pounds worth of damage.

▶ Make a list of 4 problems caused by rust.

Many companies spend lots of money trying to stop rusting.
To know how to stop rusting, we must know what causes it.

Kris and Becky investigated the conditions needed for an iron nail to rust.
They set up 3 test-tubes like the ones opposite:
They left them for a few days.

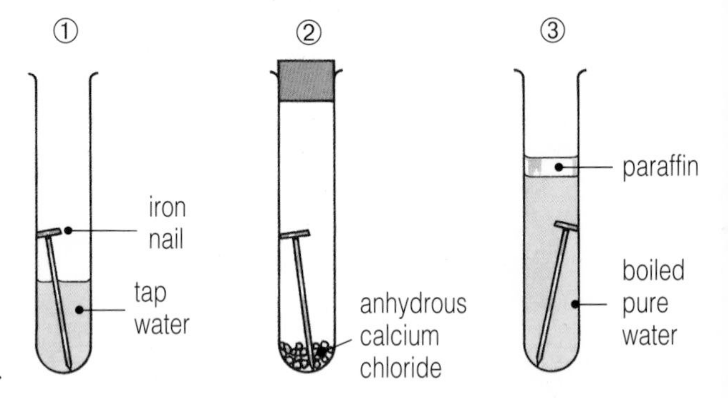

▶ Copy the table. This shows the conditions inside the tubes.

Condition	Tube ①	Tube ②	Tube ③
air	✓	✓	✗
water	✓	✗	✓

. . . after a few days

Look at the conditions in tube ②.

b What do you think anhydrous calcium chloride does**?**

Look at the conditions in tube ③.

c What do you think happens when the water boils**?**

d Why is paraffin put on top of the boiled water**?**

After a few days, Kris and Becky checked the tubes.

e What 2 substances must be present for iron to rust**?**

Have you noticed any rusting at the seaside**?**

f Rusting happens faster by the sea. Why do you think this is**?**

Rusting is an **oxidation** reaction. The iron is **oxidised**.
This means it reacts with oxygen in the air.

g Complete the word equation:

iron + oxygen (with water) ⟶ . . . .

h What is the chemical name for rust**?**

Investigating rusting

The teacher found the group's results very interesting.
She wanted Kris and Becky to do a more detailed investigation.
She didn't give them any firm ideas, but set them a question:

Imagine you are Kris. Plan an investigation into the rusting of iron.
(If you need help with this, ask your teacher for a clue!)

Stopping the rot!

To stop iron from rusting we must protect it from air and water.
Here are some ways of doing this:

- **Painting**
  This is used to protect cars and large structures such as bridges.

- **Greasing or oiling**
  This is used on moving parts of machines.

- **Plating**
  This is a thin coating of another metal which doesn't rust.
  Chromium plating is common. It gives an attractive, shiny effect.
  Zinc plating is also used. If the metal coating on the iron is zinc, the iron is said to be **galvanised**. Zinc will protect the iron even if it gets scratched.

The beauty of gold

Some metals tarnish very slowly.

Compare gold with iron.
Why does gold stay shiny and untarnished whilst iron rusts?
Write down your ideas.

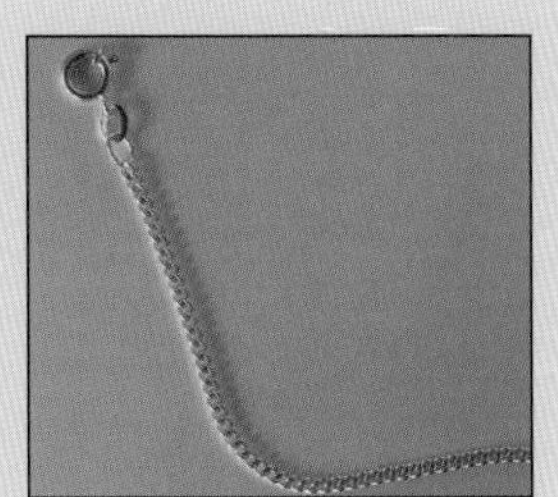

Even very old gold has a shine

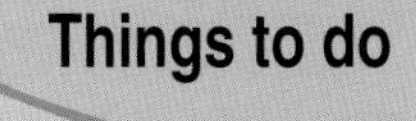

1 Copy and complete:
a) . . . . and . . . . are needed for iron to rust.
b) . . . . speeds up the rusting of iron.
c) The chemical name for rust is . . . .
d) Two methods to prevent rusting are . . . . and . . . .

2 Which rust prevention method would you use to:
a) protect a lawnmower in the winter?
b) protect a car bumper?
c) protect a school's iron gates?
Explain your choice in each case.

3 Do you think rusting happens to the same extent in all parts of the world?
Explain your ideas about this.

4 Most iron gets turned into steel.
Steel is used more than any other metal.
Find out about different types of steel.
Write about some uses of steels.
You could draw pictures too.

How reactive?

Learn about:
- metals reacting with water
- metals reacting with acid
- finding an order of reactivity

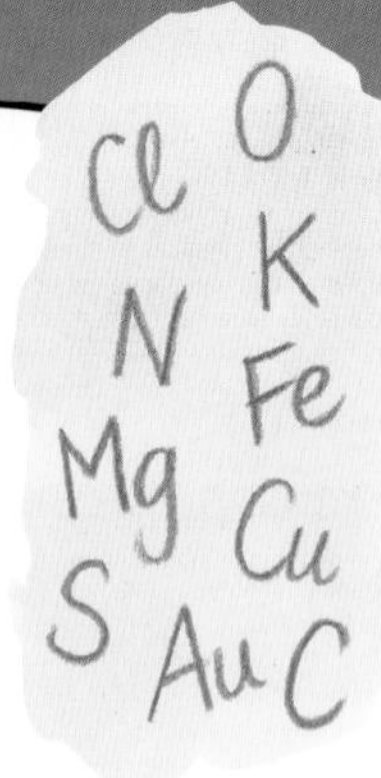

▶ Look at the symbols opposite:
Pick out the 5 which are ***metals***.
Write down their names.
Write one use for each of these metals or their compounds.

In 9E you looked at the **physical properties** of metals (for example, strength and hardness).
The **chemical properties** of metals are also important.
Does the metal **react** easily**?**

The use of a metal depends on its physical ***and*** chemical properties.
Which of your 5 metals do you think is the most reactive**?**
Why do you think this**?**

In the next experiment your teacher will show you some metals reacting.
This should help you make an order of reactivity.

Metals and water

Watch your teacher put pieces of copper, zinc, iron and magnesium in water.

a Do any of the metals show signs of reacting with the water**?**

Now watch your teacher demonstrate the reactions of lithium, sodium and potassium metals with water.

b Make a list of the safety precautions taken.
c How did your teacher cut the metals**?**
d What did the freshly cut surfaces of the metals look like**?**
e Why are the metals stored under oil**?**
f What gas is given off when the metals react with water**?**
g What type of solution is left after each reaction**?**

The word equation for lithium reacting with water is:

lithium + water ⟶ lithium hydroxide + hydrogen

h Write the word equations for sodium and potassium with water.
i Put the 3 metals into an order of reactivity.

more difficult . . .
Try to write symbols and formulae for the reactants and products in these reactions.

Metals and oxygen

When metals react with oxygen they make new substances.
These are called **oxides**.

▶ Using results from the last experiment, predict the differences you would expect if:

sodium, then iron, were heated in air;
potassium, then copper, were heated in air.

Write word equations for the reactions you predict.

Gold is unreactive. Why don't we use gold to make bridges?

Metals do not all react in the same way with oxygen or with water.

Some are ***very reactive*** – potassium.
Some are ***reactive*** – magnesium.
Some are ***unreactive*** – gold.

Metals can be put in an **order of reactivity**.
The most reactive ones are at the top of the list.
The least reactive are at the bottom.

What's the order?

Plan an investigation to produce an order of reactivity for metals.
You can use any equipment you need.
Remember to always use very small amounts of chemicals when investigating.
The chemicals you can use are:

- *metal samples* – zinc, tin, magnesium, iron, copper,
- bottles of distilled (pure) water,
- bottles of dilute acid.

⚠ eye protection

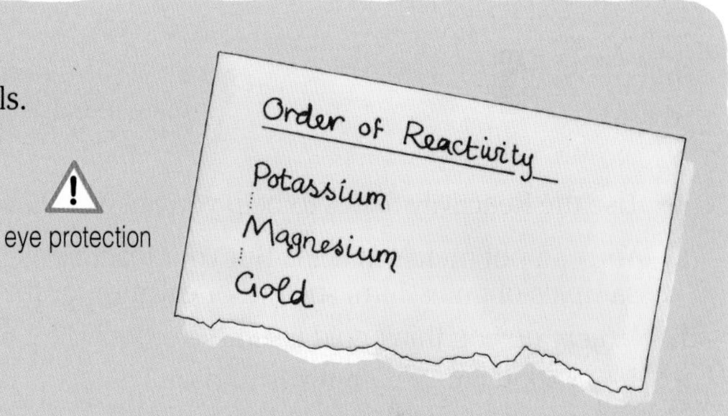

You ***must*** have your plan checked by your teacher.
Then do the investigation.
Write a report of your findings. Include your order of reactivity in your report. Are you ***sure*** your order is correct?
How could you improve this investigation?
Write down your ideas to make it better.

Things to do

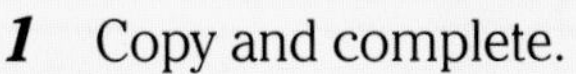

1 Copy and complete.
Lithium, ... and potassium are soft metals that . . . . with cold water.
As they react, ... gas is given off and an . . . . solution is formed.
Of the 3 metals, ... is most reactive and . . . . is least reactive.

2 Rubidium is in the same group of the Periodic Table as lithium, sodium and potassium.
Write a word equation for rubidium's reaction with water.

3 Copy and complete these word equations:
a) calcium + water ⟶
b) magnesium + oxygen ⟶
c) magnesium + sulphuric acid ⟶

4 Amy put some metals, A, B, C and D, in water. Look at the times taken for the metals to react completely with the water:

Metal	*Time (seconds)*
A	15
B	35
C	5
D	no reaction

a) Which is the most reactive metal?
b) Which metal could be copper?
c) Which metal is likely to be stored under oil?
d) How could Amy have made sure this was a fair test?
e) How could she have made her results more reliable?

Predicting reactions

Learn about:
- the Reactivity Series
- displacement reactions
- predicting reactions

You have seen that metals can be put in order of reactivity.
This is called the **Reactivity Series**. It's a kind of League Table for metals. These tests are used to find the order.

metal + air | metal + water | metal + acid

The Reactivity Series

potassium	most reactive ↑
sodium	
calcium	
magnesium	
zinc	
iron	
tin	
copper	least reactive

Can you make up your own rhymes to help you remember the order?
For example,
Please **S**top
Calling **M**y
Zebra **I**n
The **C**lass

▶ Use the Reactivity Series to help you answer these questions:

a You can put metals in acid. But your teacher will ***not*** give you samples of potassium, sodium or calcium for this. Why not?

b Where do you think gold fits in the order?

c Iron reacts slowly with water and air.
What substance is made in this reaction?

d Copper does not react easily with air or water. Would it be a good idea to make cars from copper?

There is another way of finding out an order of reactivity for metals.
We can set up ***competitions***. Competitions for oxygen are easy to do.

The big fight!

An experiment to heat magnesium oxide with copper is very boring.
Nothing happens! There is no reaction.

Heating magnesium with copper oxide is much more exciting!
There is a big reaction.

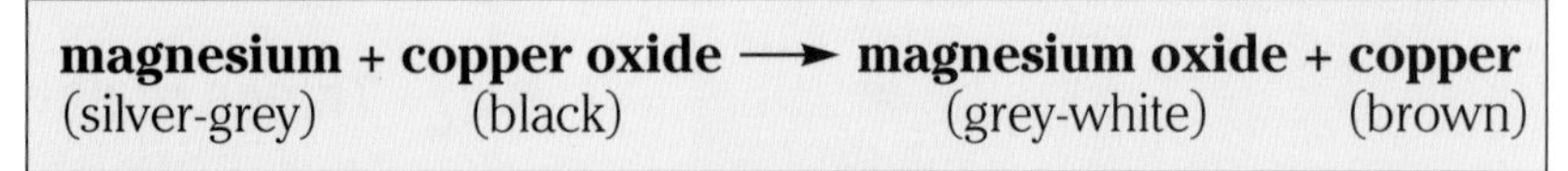

magnesium + copper oxide ⟶ magnesium oxide + copper
(silver-grey) (black) (grey-white) (brown)

This is because magnesium is more reactive than copper.
Magnesium wins the fight for the oxygen.

Reactions like this are called **displacement reactions**.
The magnesium **displaces** the copper. It pushes the copper out.
It wins the oxygen.

$$\mathbf{Mg + CuO \longrightarrow MgO + Cu}$$

Railway workers use the reaction between aluminium and iron oxide to weld tracks together. The molten iron formed is directed into the gap between two pieces of track.

Displacing metals

Try some other displacement experiments. See if you can spot reactions taking place. You should look to see if:

- a gas is made,
- any solids or solutions change colour,
- any solids disappear (dissolve) in solutions.

Take a spotting tray. Put ***small*** pieces of the 4 metals in the rows of the tray.
Use a teat pipette to add 4 different solutions to the 4 metals.
Check that yours looks like this:

Metals
- zinc
- iron
- magnesium
- copper

Solutions
- copper sulphate
- magnesium sulphate
- iron sulphate
- zinc sulphate

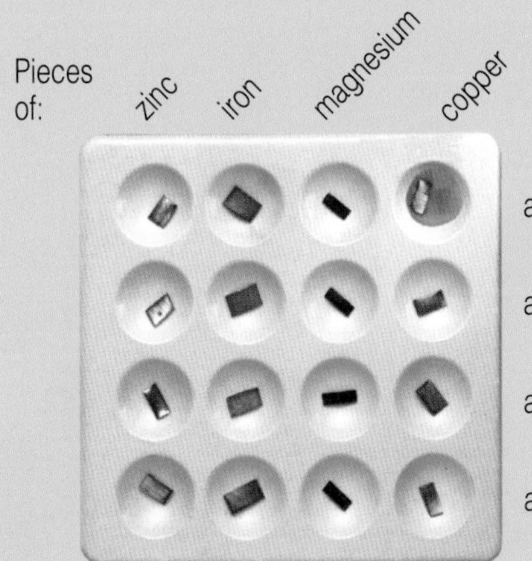

add copper sulphate solution

add magnesium sulphate solution

add iron sulphate solution

add zinc sulphate solution

Now you have added each solution to each metal.
Have there been any reactions**?**

Yes ✓ No ✗ 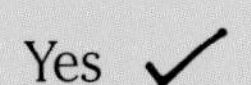

Fill in the table with ticks or crosses:

	zinc	iron	magnesium	copper
copper sulphate				✗
magnesium sulphate			✗	
iron sulphate		✗		
zinc sulphate	✗			

e Write a word equation for each reaction.

f Which of these rules is the correct one**?**
i) Less reactive metals displace reactive ones.
ii) Reactive metals displace less reactive ones.
Copy out the correct rule.

Things to do

1 Copy and complete:
In a reaction, a metal high in the Reactivity Series . . . . one below it. For example, ... could displace iron in a reaction.

2 Predict whether reactions will take place between these substances:
a) copper + zinc sulphate
b) iron + copper oxide
c) magnesium + iron nitrate
d) iron + potassium chloride
e) tin + magnesium oxide
f) zinc + copper oxide.

3 The metal nickel does not react with iron oxide. Nickel reacts with copper oxide.
a) Copy and complete:
nickel + copper oxide → . . . . + ...
b) Explain why nickel won't react with iron oxide.
c) Alongside which metal would you put nickel in the Reactivity Series?

4 Carbon is an important non-metal. How could you put it in its right position in the Reactivity Series? What experiments could you do?

Use and reactivity

Learn about:
- links between reactivity and use
- measuring reactivity

▶ How much can you remember about sodium metal?

You wouldn't keep vinegar in a container made of sodium.
Make a list of some of the reasons why not.
(Think about the properties and reactivity of sodium.)

Thinking about metals and uses

In your group discuss your answers to the questions below.
Write down your best ideas.
Use books or ROMs to help you.

a Why isn't sodium used to make knives?

b Why don't we use magnesium to make car bodies?

c Why has so much gold jewellery survived from ancient civilizations?

d Why was bronze used before iron?

e Why was iron used before aluminium?

f Which metals are found naturally?

g The metal magnesium is not found naturally.
What are the main sources of magnesium? How is it extracted?

h How are metals recycled?

i What factors affect how expensive a metal is?

From around 1800 BC people began to make bronze.
They mixed copper and tin over hot fires.

Look at your answers to the questions above.
Try to spot links between a metal's reactivity and

- whether it is found naturally,
- when it was first used,
- how it is extracted,
- how it is used nowadays.

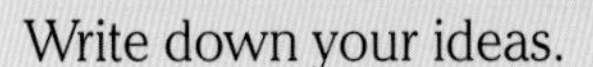

Write down your ideas.

Alloys

An **alloy** is a mixture of metals. The mixture has different properties to the original metals.
This may make it more suitable for some uses.

Brass is a mixture of copper and zinc.
It is used to make musical instruments.

j What properties does brass have?

Solder is made from lead and tin.

k How does solder join up electrical circuits?

Differences in reactivity

In this experiment you will be looking more closely at 3 metals – zinc, magnesium and aluminium.

You will be investigating the differences in their reactivity with hydrochloric acid.

You need to compare the reactivities very carefully.

Try to think of observations and measurements you could make.
How could you make a graph of your results?
Which variables do you need to control?
How could you do this?
What safety precautions are needed?

Write a plan for your investigation.
Show your plan to your teacher.
Then carry out your investigation.

Try to explain your results using your scientific knowledge and understanding.

Put the 3 metals in an order of reactivity.
Now check this order with a copy of the Reactivity Series.
What do you notice?

Which of these 3 metals do you think is the most useful?
Give reasons for your choice.

You may need some of this equipment:

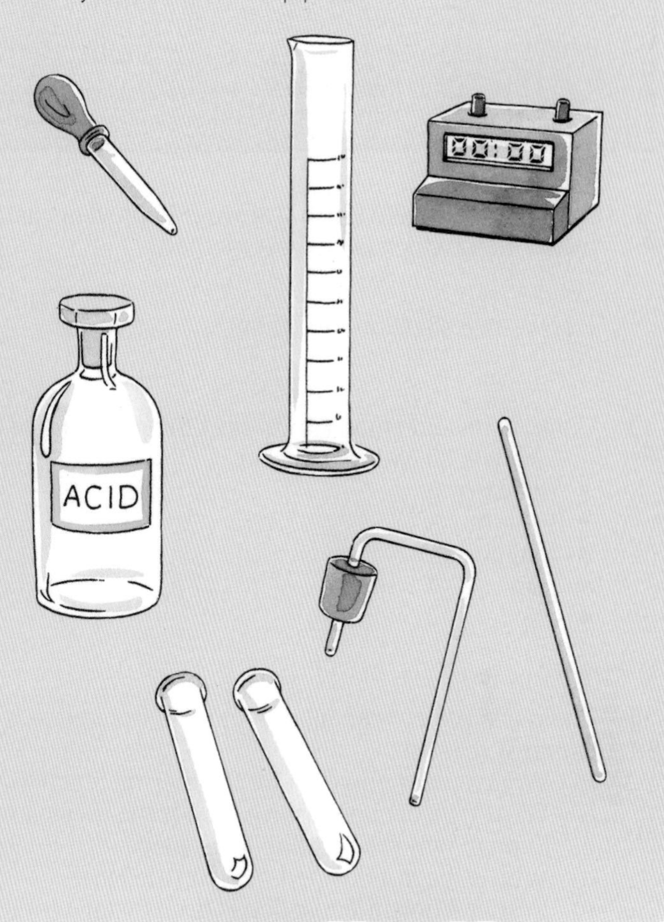

Things to do

1 Copy and complete using the words in the box:

reactive gold high unreactive

a) Metals found naturally are usually . . . .
b) Metals . . . . in the Reactivity Series usually have to be extracted.
c) Some very . . . . metals need electricity to extract them.
d) Unreactive metals such as . . . . tarnish very slowly.

2 Make your own summary sheet of the reactions of metals.
Include information about the Reactivity Series.
You could draw cartoons or diagrams to help you remember the reactions.

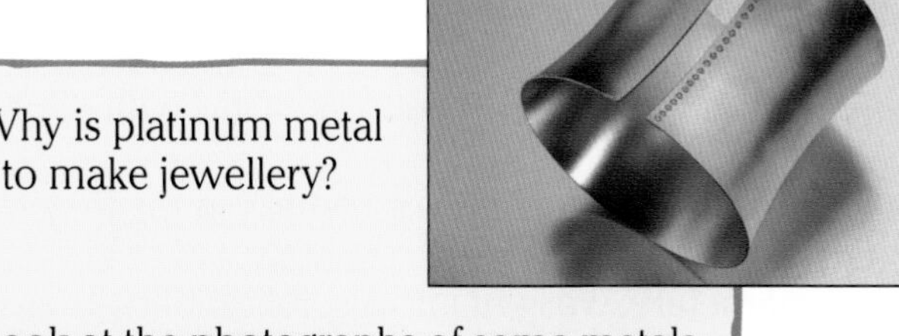

3 Why is platinum metal used to make jewellery?

4 Look at the photographs of some metals reacting with dilute hydrochloric acid.

magnesium

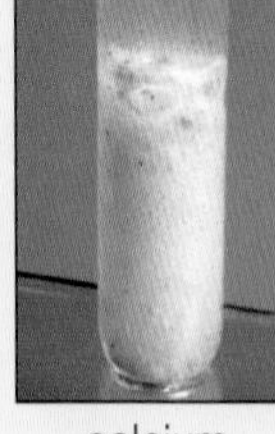
calcium

copper

a) Use this reaction to put these metals into a Reactivity Series. Put the most reactive at the top.
b) Write a word equation for the reaction of the acid with magnesium.
c) What is the name of the gas made in this reaction?

Questions

1 Lee and Asha were looking at ways to prevent rust. Their teacher asked them which is the best method.

Who is right?
Plan an investigation to find out.

2 Construct a table that summarises the reactions of the following metals with air, water and acid:
- potassium
- sodium
- magnesium
- iron
- copper.

3 The reaction between aluminium and iron oxide is called the **thermit** reaction.
a) Write a word equation for the thermit reaction.
b) What type of chemical reaction is the thermit reaction?
c) Explain why the reaction takes place.
d) Why is the iron formed molten (melted)?
e) Describe how the reaction is used by rail workers.

4 Imagine that a new metal, X, has been discovered. Investigations into the chemical properties of X have found that the metal does not react with cold water. It does react slightly with steam and slowly gives off bubbles of gas with dilute hydrochloric acid.
a) Explain where you would place metal X in the Reactivity Series.
b) A piece of metal X was put into a solution of copper sulphate. Explain what you would expect to happen (include a word equation).
c) Write a word equation for the reaction of metal X with dilute hydrochloric acid.
d) A company is thinking of using metal X to make food cans. What things should scientists consider before going ahead with the idea?

5 In the Reactivity Series, carbon is usually placed just above iron. Carbon and copper oxide are both black powders. Unfortunately the labels have come off their containers in the laboratory. What experiments could you do to find out which powder is which?

Environmental chemistry

Our environment is precious and we should all do our best to protect it. In this unit you can learn about different sources of pollution and how scientists have a major role to play in controlling and monitoring environmental conditions.

pH in the garden

Learn about:
- pH of soils
- reducing soil acidity

pH of soil

The pH of a soil is very important. Some plants grow well in acidic soil. Some would grow better in neutral or alkaline soil.

Gardeners and farmers need to know the pH of their soil. The soil gets too acidic sometimes. This can stop plants growing well. The farmers could add lime (an alkali) to change the pH of the soil.

It's a fact!

You can buy pH test kits from garden centres to test your own soil.

Testing soils

Your teacher will give you some different soils to test.

Take your first soil sample and put 2 spatula measures in a test-tube. Add about 5 cm^3 of distilled water. Stopper the tube and shake it for about a minute.

Set up a filter funnel and paper. Filter the soil mixture into another test-tube. Add universal indicator to the filtrate. Record the pH value.

Repeat this test with other soils.

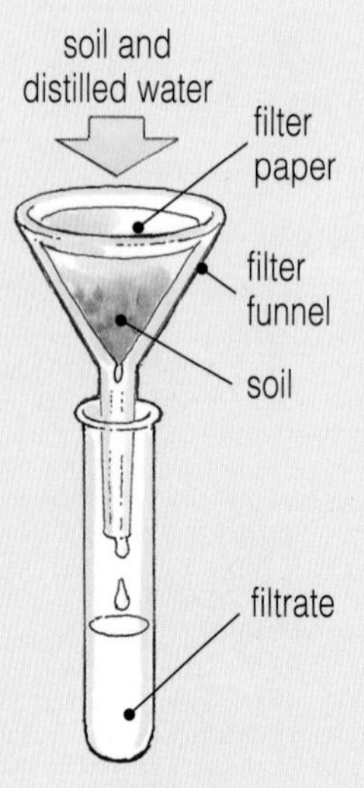

Soil sample	*pH value*
A	
B	
.	
.	
.	

Write down answers to the following questions.

a Why do you use ***distilled*** water in the test?

b Why do you 'shake for about a minute'?

c Put your soils in order, with the most acidic first.

Look at the plant pH preference list (opposite) from the pH test kit:

d Which of your soil samples do you think could grow (i) apples? (ii) potatoes? (iii) blackcurrants?

e Which crops could soil A grow?

Find some other plants that prefer different types of soil. Record your findings in an information sheet for a garden centre.

Plant pH preference

Name	*pH preference*
apple	5.0–6.5
potato	4.5–6.0
blackcurrant	6.0–8.0
mint	7.0–8.0
onion	6.0–7.0
strawberry	5.0–7.0
lettuce	6.0–7.0

Neutralising acidic soil

Collect 2 spatula measures of an acidic soil from the last experiment. Mix in a spatula of powdered limestone, which contains calcium carbonate.
Repeat the method from the last experiment to find the pH of the soil mixed with limestone.

f How has the pH value of the soil changed once you have added limestone to the acidic soil?

Repeat the method above, but this time add powdered lime (calcium hydroxide) to the acidic soil.

g Which method was better at neutralising the acidic soil?

h What would a farmer do to decrease the pH value of an alkaline soil?

A farmer neutralising acidic soil

Different types of soil

As well as pH, soils can differ in other ways.

They vary according to:

- the size of the rock fragments they contain;
- the chemical compounds that make up the rock fragments;
- the amount of matter from dead plants and animals (called **humus**) in the soil.

This sandy soil contains relatively large rock fragments so water drains well through it

Things to do

1 Make a drawing to show which plant would grow best in each pot.

2 Lime has a pH value of about 9.
Citric acid has a pH value of about 4.
Which should you add to a neutral soil to grow apples?

3 Clay soils have very fine particles.
Do you think they let water drain through them easily?
Explain your answer.

4 Plan an investigation to see how the pH of soil depends on the amount of lime added to it.

5 Fertilisers are used to help plants grow.
They give nutrients (food) to the soil and can change its pH value.
Do you think we should use fertilisers?
Why? Why not?

Water for life

Learn about:
- the importance of water
- testing water

Water is a **compound**. It is made from 2 elements.
Can you remember which elements combine to make water?

Water has different states.

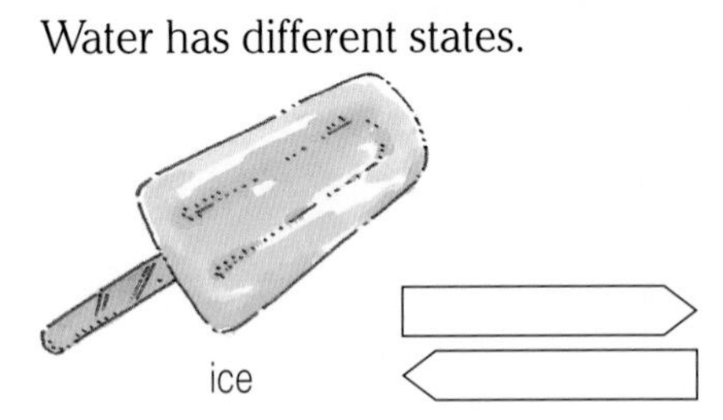

▶ Copy out these states. Add the following words in the right places:

evaporates ***freezes*** ***condenses*** ***melts***

When water changes its state, it is a **physical change**.
In a physical change no new substances are made.
You can also reverse the change quite easily.

▶ Your teacher will show you what ***1 litre*** of water looks like.
How much water does your family use each day?
Estimate the number of litres.
Think about the water your family uses for:

- flushing toilets
- washing people (baths, showers, . . .)
- washing clothes
- washing dishes
- cooking
- drinking.

Are there other uses?

Now do a rough calculation to estimate how much water is used by all the families of people in your class.

▶ In many other countries water is in short supply.
Why do you think this is?
Many people have to carry the water they need for miles.
How do you think this affects the amount of water they use?
(Think about carrying a bucket of water around school for a few hours . . .)

Testing water

What do you think ***pure*** water is? What is the difference between impure water and pure water?

Do the following tests on your samples of pure and impure water. Record your results in a table like this:
For each sample of water,

1 Measure its boiling point.
2 Measure its freezing point.
3 See if it conducts electricity.
4 Put a drop of the sample onto a piece of blue cobalt chloride paper. Note what you see.
5 Put a drop of the sample onto a piece of universal indicator paper. What is the pH number?

eye protection

Test	Pure water	Impure water
1. Boiling point		
2. Freezing point		
3.		

Which of the tests gives the same result for pure water and impure water?

Investigating rain water and stream water

Plan an investigation to find out whether rain water or stream water is purer.
- What do you need to measure?
- How will you make sure it is a fair test?
- How will you record your results?

Check your plan with your teacher who may then let you carry it out.

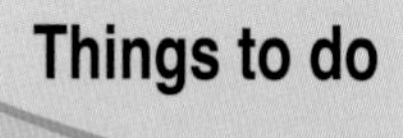

1 Copy and complete the following sentences:
a) The melting point of pure water is . . . . °C.
b) The boiling point of pure water is . . . . °C.
c) When . . . . melts it forms liquid water.
d) When liquid water . . . . it forms steam.
e) The pH of pure water is . . . .
f) A change of state is a . . . . change.

2 Draw diagrams to show the arrangement of particles in the 3 states of water.

3 Design a poster to encourage people to use less water.
Why might it be important to save water?

4 Imagine that you have the job of selling water. Think about its properties and possible uses. Make an advertising leaflet to help sell the water.

5 FIZZO is a popular new blackcurrant flavoured drink. One of its ingredients is water.
a) Describe an experiment to get pure water from the drink.
b) Describe an experiment to find out whether the purple colouring is a pure, single substance or a mixture.

The water cycle

Learn about:
- the water cycle
- cleaning water
- fluoride in water

What states of water can you see in each of the pictures?

▶

Do you agree with Jack?
Write about 4 or 5 lines to explain why.

We use billions of litres of water every day.
Have you ever wondered why we don't run out of it?
The answer is that it is **recycled**.

The water cycle is a very important process in nature.

In your group make a poster to show the water cycle.
This picture could be the ***start*** of your poster.

The labels give information about some parts of the water cycle.
Add them to your picture in the right places.
Where is the rain likely to fall?
Draw it on your picture.
Make sure you put arrows on your picture to show which way the water is moving.

Labels
- water evaporates from the sea
- water as a gas (vapour) moves upwards
- as it gets colder, water condenses into droplets to form clouds
- wind moves clouds
- droplets get heavier, and water falls as rain, hail or snow
- the water returns to the sea through lakes and rivers

a Energy from the Sun is very important to the water cycle. Which parts of the cycle does this affect?

b Why does water sometimes fall as ***snow*** rather than rain?

c What might cause impurities in the rain?

Before water goes back to the sea it is piped into our homes from lakes, rivers or from underground.
You couldn't drink water which came straight from rivers.

Why not**?**

All water must be cleaned before we can use it.
This happens at a water treatment plant (waterworks).
Very large amounts of water must be cleaned cheaply.
The process used is called **filtration**.

Would you like to drink this?

Cleaning water

Your teacher will give you some muddy water.
Your task is to get the cleanest water you can from the muddy water.
You can use any of the equipment your teacher will give you.

Draw a diagram of the arrangement you use to get the clean water.

You must not drink this water. Why not**?**

Find out how water is cleaned at a water treatment plant.

▶ In some areas sodium fluoride (NaF) is added to the water after it has been cleaned. Dentists think that this **fluoride** helps to stop tooth decay. However too much fluoride can be poisonous.
Carry out some research into the fluoridation of water supplies.
Gather information for and against adding fluoride.
Think about any bias in the evidence presented by the different authors.
Explain why you think some information might be biased.

d Do you think fluoride should be added**?** Explain your views.

e Suggest other ways of reducing the decay of your teeth.

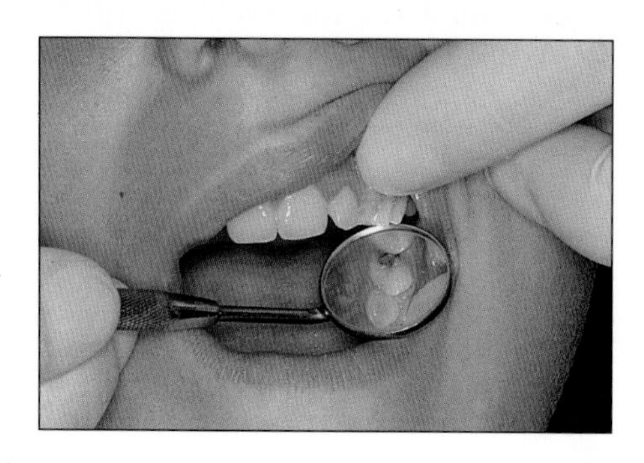

Things to do

1 Imagine you are a water particle in a drop of rain. Write about your adventures as you pass through the water cycle. Finish your story when you reach the clouds again. Be sure to write about changes of state.

2 Draw a diagram of the apparatus used normally for filtration in the laboratory. Label the **residue** and **filtrate** on your diagram.

3 Design a piece of apparatus to measure rainfall. Your apparatus should be able to remain outdoors for long periods of time.

4 Will it rain for 40 days after a rainy St. Swithin's Day? There are lots of sayings about the weather. Use books and ask friends or relatives about these sayings.
Write down as many as you can.
Do you think there is truth in any of these?

5 Find out about **hard water**.
What is the difference between **hard water** and **soft water**?
Give one advantage and one disadvantage of living in a hard water area.

Water problems

Learn about:
- causes of acid rain
- effects of acid rain
- reducing acid rain

▶ Think about countries all over the world.
Make a list of the problems that can happen when there is:

a too much rain,

b too little rain.

The water cycle shows us that we need rain.
Do you think our rain water is pure**?**

▶ Look at the pictures below. In your group discuss how you think these things affect rain water.

gases contain carbon dioxide and oxides of nitrogen

gases contain carbon dioxide and sulphur dioxide

Testing pH

Your teacher will give you some solutions.
Use universal indicator to measure their pH values.
Put the solutions in order from the least acidic to the most acidic.

Rain water is usually acidic.
This is due to acidic gases in the atmosphere.
The air around us contains about 0.04% carbon dioxide (CO_2).
We breathe out carbon dioxide naturally and plants use it during photosynthesis.
But the amount of CO_2 in the air is building up. This is because we are using lots of fossil fuels which give off CO_2. This could eventually cause more water problems. Do you know why**?** (See 9G5.)

Remember,
– pH1 is strong acid.
– pH14 is strong alkali.

Dissolved CO_2 causes rain to have pH5 or 6.
But other gases are a bigger problem.
Oxides of sulphur and nitrogen make the rain much more acidic.
The pH can fall to 4 or even lower.

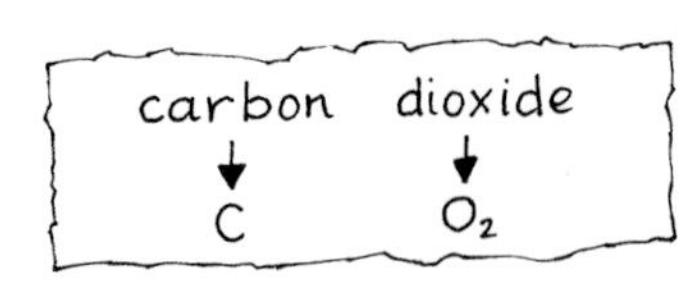

What are the effects of acid rain?

Use the solutions from the ***Testing pH*** experiment on page 86 to find out the effects of acid rain. Record all your results in a table.

- Take a metal or rock sample and put it in a test-tube.
- Add one of the solutions to it until the tube is $\frac{1}{3}$ full.
- Leave the tube in a rack.
- Write down what you see.

eye protection

Now repeat the experiment using the other solutions and the other samples of metal and rock. Leave an experiment set up for several lessons and observe the tubes over time.

What do you notice? Write 2 or 3 sentences to explain your conclusion.

As well as attacking building materials and metals, acid rain also affects living things. It damages plants, especially trees, and animals that live in rivers and lakes.

Find out more about the effects of acid rain on living things.

How can we reduce acid rain?

▶ Find out how catalytic converters and sulphur precipitators work.

Monitoring our water supply

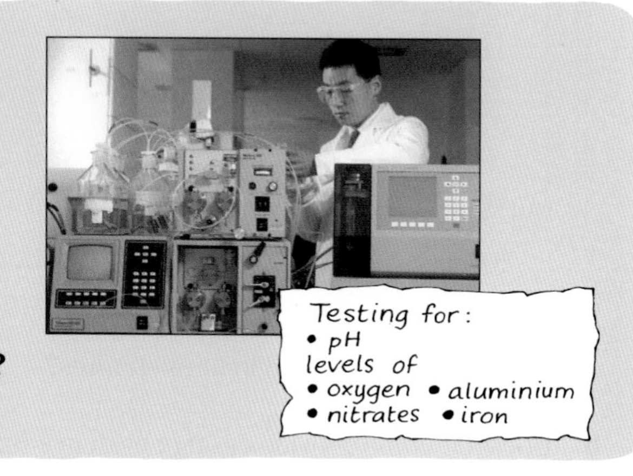

Our water supply is checked carefully to make sure it is clean and safe to drink. The water companies test samples from homes, businesses, reservoirs and treatment works.
They make sure that the levels of chemicals in the water are within set limits. For example, pH cannot be lower than 5.5.
Ideally it should be close to 8.
If acid rain causes it to have a lower pH, the water must be treated.

c What could the water company add to water to increase its pH?

d What is added to water to kill bacteria before we drink it?

Testing for:
- pH

levels of
- oxygen
- aluminium
- nitrates
- iron

Things to do

1 a) Write down 2 ways in which acid rain is made.
b) Write down 2 effects of acid rain.
c) Write down 2 ways in which acid rain can be reduced.

2 Look at the design of your school.
How is your school designed to stop problems of flooding?
Draw any important features.

3

Another drought!

We get a few days of sunshine in this country and we're told there's a drought. We've got to save water. A leaflet has come through my door to tell me what I can't do.

I'm fed up with it. I don't like showers. Why can't I have a bath if I want? Why can't I water my garden? I've spent hundreds of pounds on my plants *and* I pay for my water!

A country where it rains as much as it does here should never have water shortages. Who is mismanaging the water supply?

Ivor Grouch

Do you agree with Ivor Grouch?
Write a letter to the newspaper to support or object to Ivor Grouch's views.

Global warming

Learn about:
- the greenhouse effect
- the effects of global warming
- evaluating evidence

People are getting more and more worried about the increasing amount of carbon dioxide in our atmosphere.

All living things respire and produce carbon dioxide. However, nature has ways of absorbing carbon dioxide from air.
For example, the oceans dissolve huge amounts of the gas and plants use carbon dioxide during photosynthesis.

But these natural processes can't cope with the carbon dioxide produced as we burn more and more fossil fuels and industry spreads around the world.

Look at the graph opposite:
It has been estimated that half of the carbon dioxide released since 1750 has been produced since 1974!

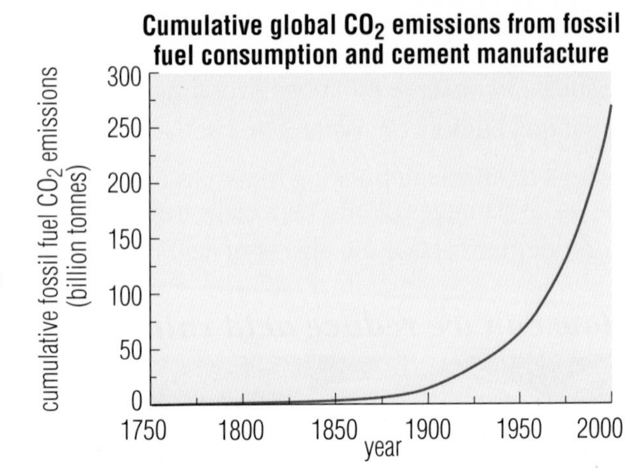

Why worry?

Look at the average global temperatures for the last 80 or so years:

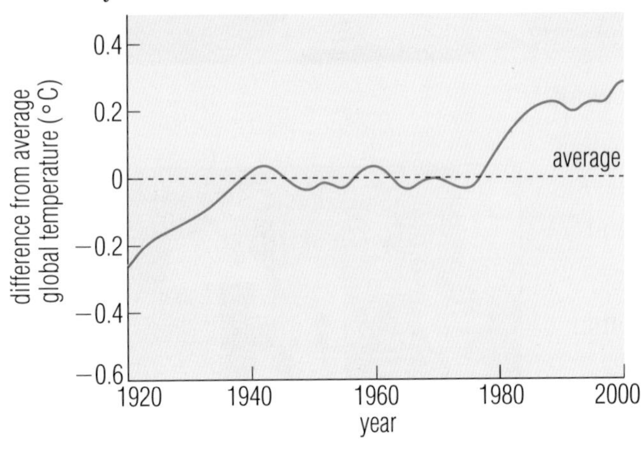

Some people blame global warming for extreme weather events that seem to be happening more often

a Is there a general trend in temperatures? What is it?

b Does the graph provide enough evidence for global warming? Explain your answer.

Many scientists are convinced that the increased level of carbon dioxide in the air is causing global warming. Carbon dioxide is a **greenhouse gas**. It stops heat escaping from the Earth as it cools down.
(Take a look at the diagram opposite.)

However, a few scientists argue that the temperature changes in the graph above are just natural variations. They argue that such changes have always happened.

c Name a period in the Earth's history when temperatures were much different to those we have today.

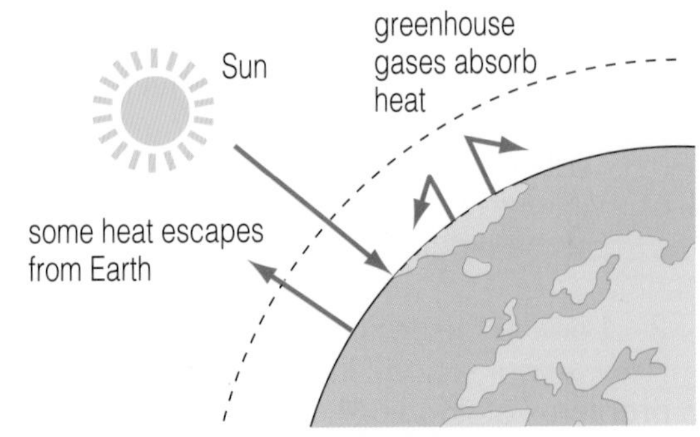

Carbon dioxide and water vapour are **greenhouse gases**

Researching the greenhouse effect

Use books, videos, ROMs and the internet to find the answers to some questions about the greenhouse effect. Here are some suggestions:

- Which gases are the main 'greenhouse gases'? Where do they come from?
- What are the possible consequences of global warming?
- Why is there still some debate about whether global warming is happening or not?
- How can we reduce the volumes of 'greenhouse gases' given off into the atmosphere?

Present your findings as an information booklet suitable for the general public.

Industry produces millions of tonnes of carbon dioxide each year.
For example, in the extraction of iron from its ore, iron oxide (see 9H4):

iron oxide + carbon ⟶ iron + carbon dioxide
iron oxide + carbon monoxide ⟶ iron + carbon dioxide

Limestone is also heated as a raw material for making cement.
You can try out the reaction we get inside a limekiln. Look at the instructions below.

Roasting limestone

Take 2 limestone chips which look similar.
Heat one chip strongly in a hot Bunsen flame.
Heat it for 10 minutes. Then let the chip cool.
Compare the heated and unheated chips:

- Appearance – what do they look like?
- Do they scratch easily? Use an iron nail to test. (Don't touch the solids.)
- Add 2 drops of water to each chip and test with pH paper.
- Add one drop of dilute acid to each.

Record all your results in a table.
Do you notice any difference between the chips?

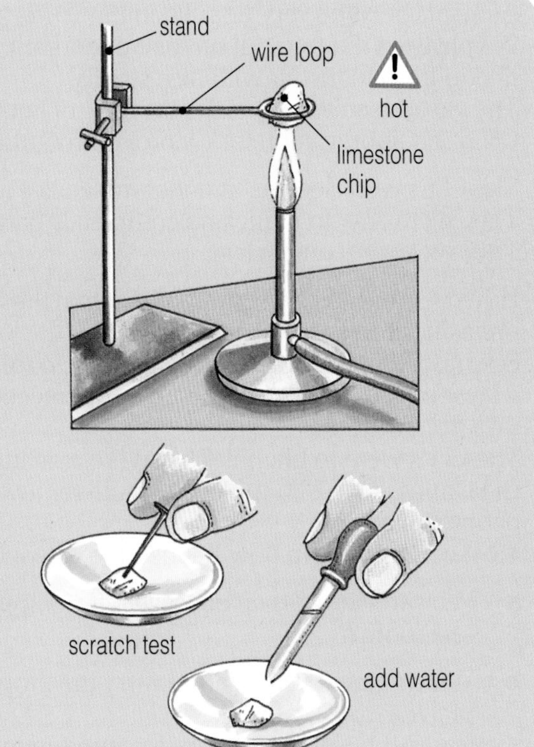

When you heated the limestone chip it decomposed (broke down).
Two products were made in the reaction:

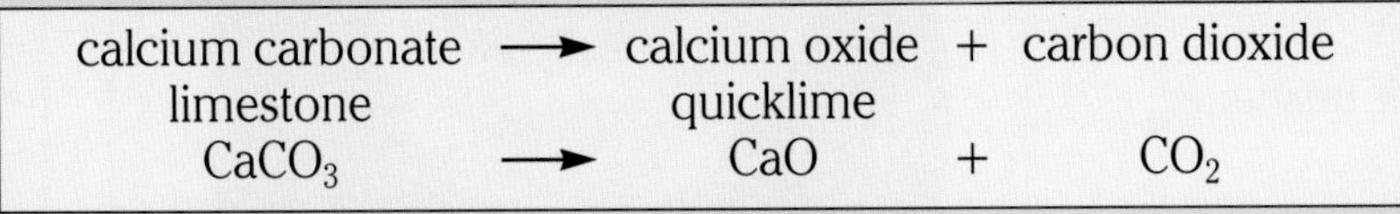

calcium carbonate	⟶	calcium oxide	+	carbon dioxide
limestone		quicklime		
$CaCO_3$	⟶	CaO	+	CO_2

d What happened to the carbon dioxide?

e Quicklime is an alkali. Sometimes farmers use it on soil. Why?

Things to do

1 Copy and complete:
In nature, ... dioxide gas is removed from the air by plants during . . . . and by the . . . . as the gas dissolves.
However, we are now producing so much of the gas that people are worried about . . . . warming because of the . . . . effect.

2 Politicians have organised global conferences to try to agree targets for reducing carbon dioxide emissions.
However, some countries have refused to cooperate.
Write a letter to the president of one such country. Explain your views on global warming.

Chemistry at Work

Learn about:
- ozone depletion
- monitoring pollution

Losing the ozone layer

An **aerosol** is a gas with tiny droplets of liquid mixed in with it.
Do you use any aerosols**?**

Aerosols became very popular in the 1970s and 1980s.
People used them to put on their deodorant, their hair-spray and to freshen the air with nice smells.
The gases used to propel the droplets of liquid were called CFCs (an abbreviation for ChloroFluoroCarbons).

The CFCs were seen as 'wonder molecules'.
They were easy to make and completely harmless to use.
They were very unreactive.

However, scientists discovered later that these CFCs were not as harmless as was first thought.
Scientists found a hole appearing in the **ozone layer**. This is a layer of gas in the atmosphere that protects us from harmful rays from the Sun.
The CFCs were to blame! They were reacting with ozone gas molecules. One CFC molecule could remove thousands of ozone molecules.
Most countries have now banned the use of CFCs.

a Why are aerosols an effective way to apply air freshener to a room**?**

b Which gases were used to carry the droplets of liquid out of the aerosol cans**?**
Name the elements in these compounds.

c Why is there a hole in the ozone layer**?**

d Find out:
i) the dangers of losing the ozone layer.
ii) the importance of ozone in the history of life on Earth.
iii) which gases we use now in aerosol cans.

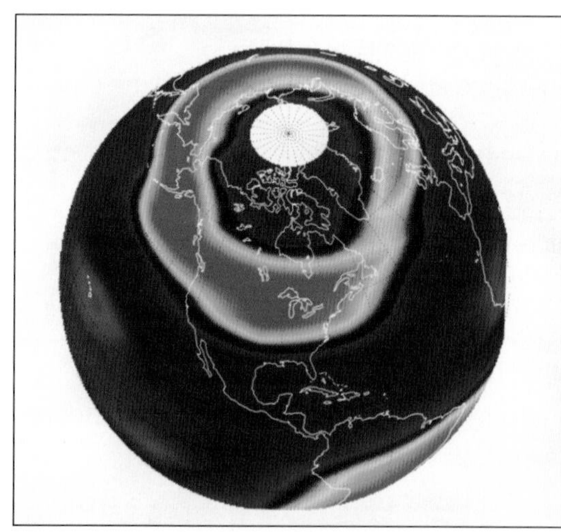

This is a special photograph taken from space.
It shows the hole in the ozone layer.

Fortunately, the latest reports from scientists monitoring the ozone layer show that the hole appears to be shrinking.
That's good news for all of us!

Monitoring pollution

Environmental scientists have been carefully monitoring levels of pollutants since the first Clean Air Act in 1956.
Since then computerised technology has given them new datalogging instruments. Instead of taking samples then doing experiments and calculations back in their laboratories, continuous monitoring on-site takes place.

The latest datalogging equipment gathers readings constantly. It stores the data every few seconds, and produces averages for 15-minute and hourly intervals.

Many places make this data available to the public.

Levels of air pollutants are monitored continuously

Look at the data from Cambridge City Council below:

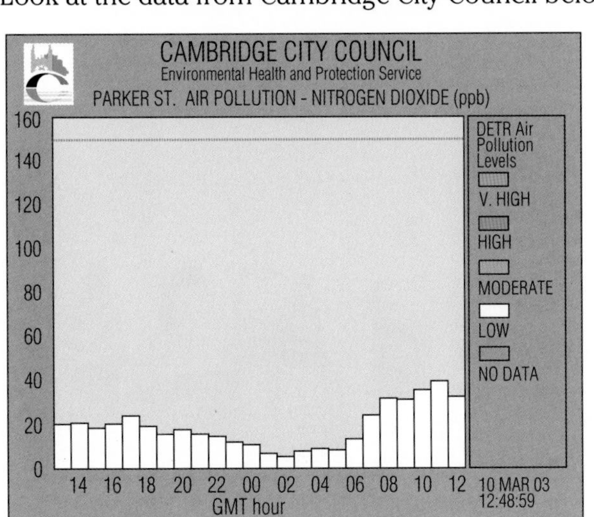

(ppb = parts per billion)

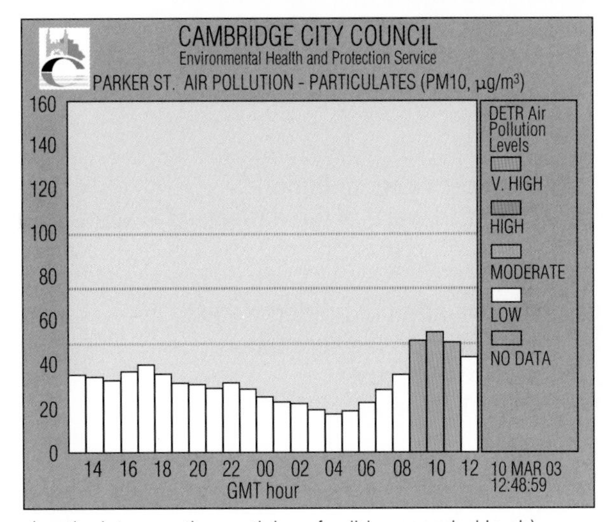

(particulates are tiny particles of solid suspended in air)

Nitrogen dioxide and particulates can both be given off by cars.
Diesel engines produce more particulates than petrol engines.

e Compare the data gathered over 24 hours in Cambridge.
Comment on the patterns in the data.

f The levels of these pollutants are classed as 'low'.
What have councils done to reduce air pollution?

Burning coal is responsible for much of the sulphur dioxide that causes acid rain.
Look at the data that shows black smoke and sulphur emissions since 1962:

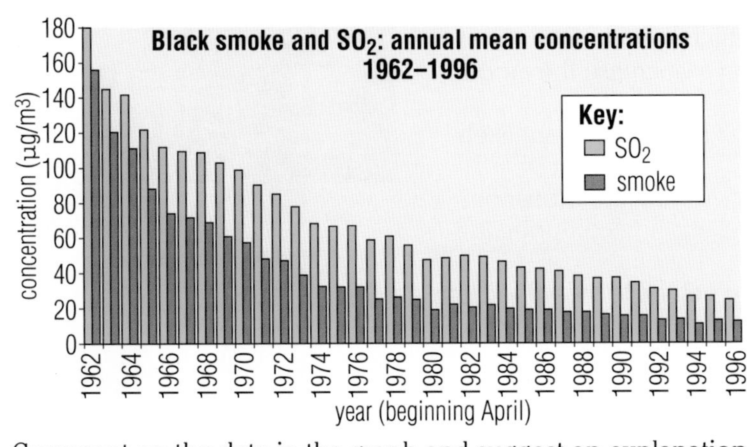

g Comment on the data in the graph and suggest an explanation.

Questions

1 Every year, 250 tonnes of lead from fishermen's weights gets into the environment.
It is thought that birds like the mute swan eat the lead weights when feeding.

a) How do you think the lead gets into the blood of the swan?

Swans in town areas are more affected by this type of poisoning than those in the country.

b) Why do you think this is?

c) How could this pollution be reduced without banning fishing altogether?

2 What can cause water pollution?
Make a poster to warn of the dangers of water pollution.

3 Do a survey of different bottled waters.
What substances do they contain?
How much do they cost?
Why do people buy bottled water?

4 Suppose each person in your family uses 200 litres of water each day.

a) Work out how much water your family uses each day.

b) Work out how much water your family uses each year.

c) Find out how much your family pays each year for its water supply. (Why do we have to ***pay*** for water?)

d) How much does your water cost per litre?

e) Do you think this is cheap or expensive? Explain your view.

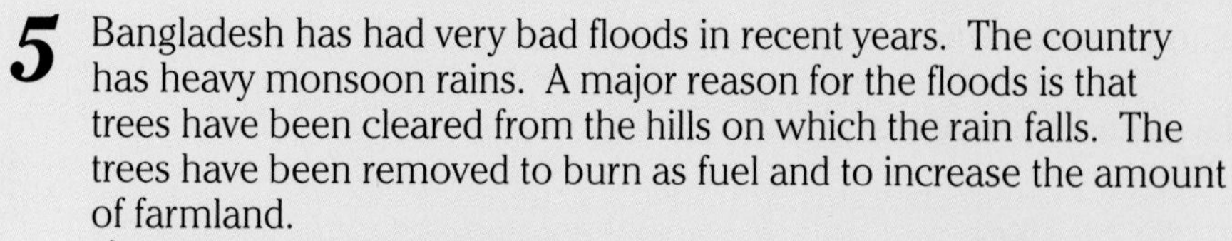

5 Bangladesh has had very bad floods in recent years. The country has heavy monsoon rains. A major reason for the floods is that trees have been cleared from the hills on which the rain falls. The trees have been removed to burn as fuel and to increase the amount of farmland.

a) Use an atlas to find Bangladesh. Name 2 rivers which are likely to overflow near Dhaka.

b) Why does clearing the trees cause floods during the monsoon?

c) What sort of help do the people of Bangladesh need during the floods?

d) How could the people of Bangladesh be helped to stop floods in the future?

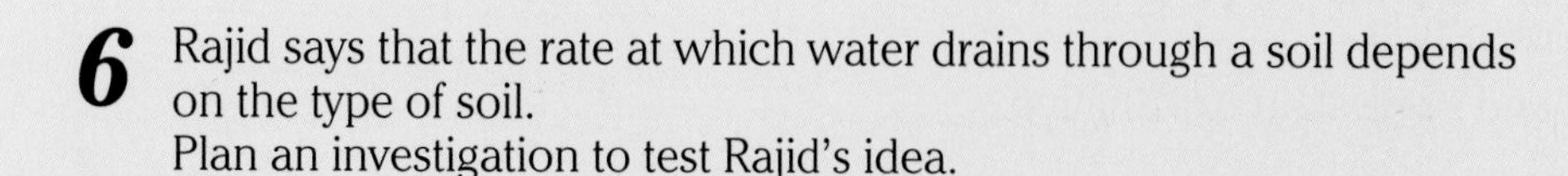

6 Rajid says that the rate at which water drains through a soil depends on the type of soil.
Plan an investigation to test Rajid's idea.

Using chemistry

Chemical reactions keep you alive.
Today you'll probably be eating food made by chemical reactions.
Reactions may also be keeping you warm.

Some reactions are so powerful they can put a rocket into space . . . .

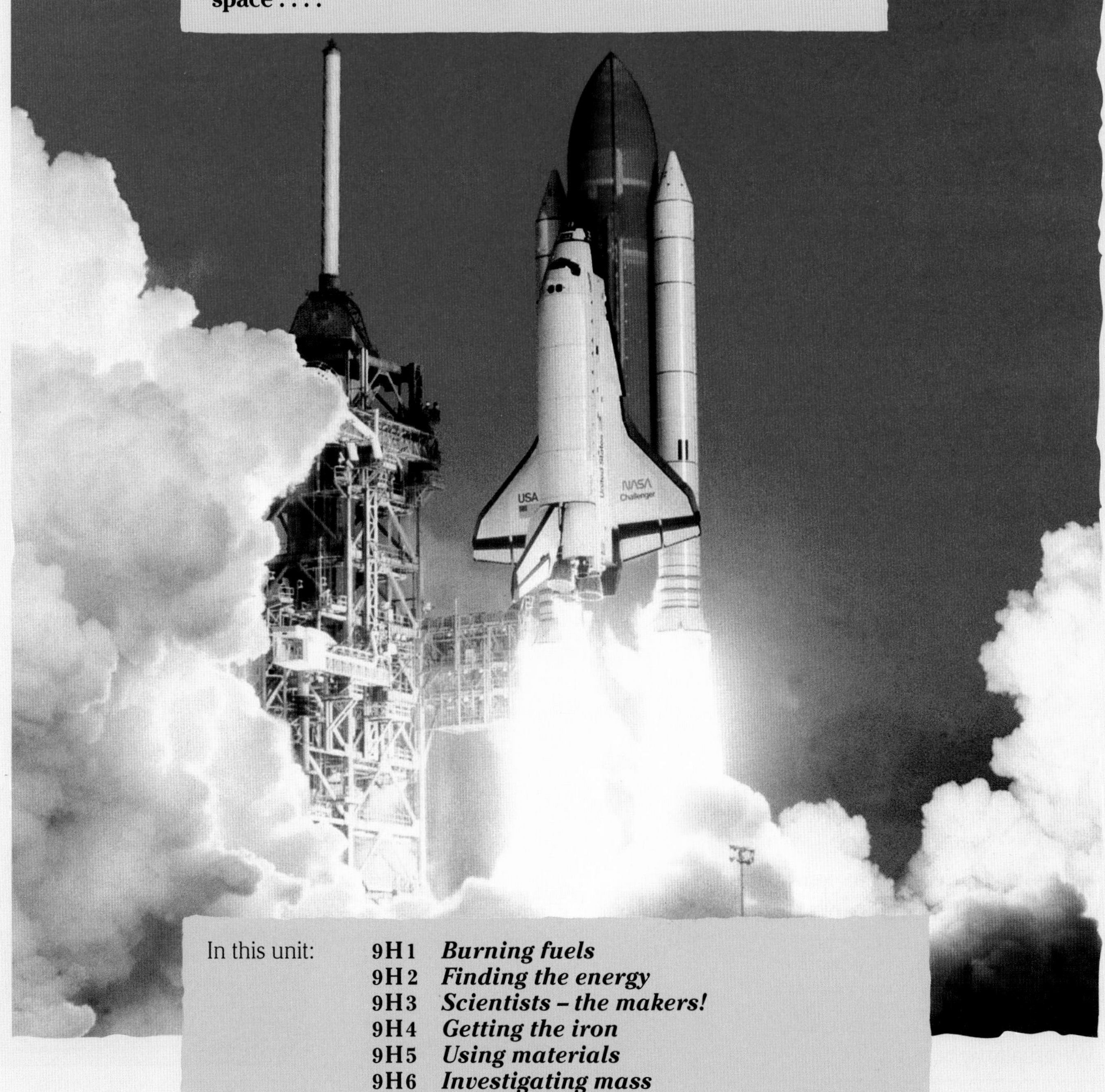

Burning fuels

Learn about:
- fuels
- incomplete combustion

Look at the ideas opposite:

a Why does Josh think burning is a useful reaction?

b Why does Emma think that burning fuels is spoiling our planet?

Remind yourself of the work you did on burning in Book 7.
Your teacher might show you the experiment below.

Burning fuels

to pump

A

ice

lime water

You can burn fuels containing carbon and hydrogen.

c What do you see in tube A?

d How can you test if this is water?

e What happens to the lime water?

If there is enough oxygen, the fuel makes carbon dioxide (CO_2) and water (H_2O).

This is called **complete combustion**.

f What else do you notice inside the upturned funnel? What substance do you think this is? Where has it come from?

Methane is another fuel. It has the formula CH_4. When it burns and there is complete combustion, it makes carbon dioxide and water.
The balanced symbol equation is:

$$CH_4 + 2O_2 \longrightarrow CO_2 + 2H_2O$$

Incomplete combustion

Read this information from a BMW owner's handbook:

Do not run the engine inside enclosed spaces, as inhalation of the exhaust gases can lead to unconsciousness and death. They contain carbon monoxide, which is colourless and odourless, but highly toxic.

g Where does the carbon monoxide gas come from?

h What would be the hazard sign for carbon monoxide?

There is only a limited amount of oxygen inside a car's engine. Instead of all the carbon in the petrol or diesel turning into carbon dioxide (CO_2), some of the fuel produces carbon monoxide (CO) and particles of carbon (C).

This is called **incomplete combustion**.

Look at these balanced symbol equations for the incomplete combustion of methane:

$$CH_4 + O_2 \longrightarrow 2H_2O + C$$

$$CH_4 + 1\tfrac{1}{2}O_2 \longrightarrow 2H_2O + CO$$

We usually write this as:

$$2CH_4 + 3O_2 \longrightarrow 4H_2O + 2CO$$

When fuels, such as petrol, burn they release their stored chemical energy

Monitoring carbon monoxide

Look at the graph showing carbon monoxide in the air in the Shetland Islands, Scotland:

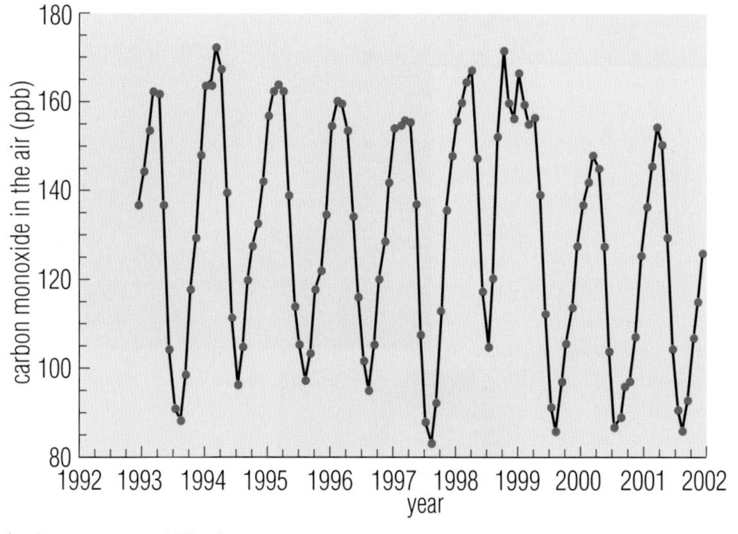

(ppb = parts per billion)

The amount of carbon monoxide in the air changes throughout the year

i What trend do you notice each year?

j Suggest an explanation for the trend.

Burning a match

Do you know why a match lights?
The 'blob' on the end of a match is called the match head.
It is made of a mixture of potassium chlorate, sulphur and carbon.

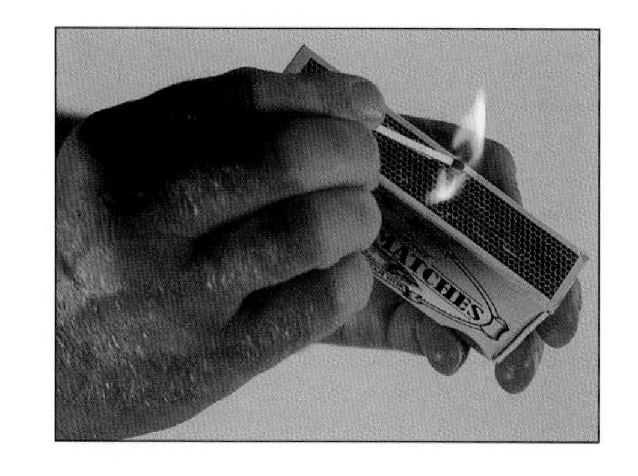

k Which 2 of these chemicals are elements?

l Are these 2 elements metals or non-metals?

m What does the 'ate' in potassium chlorate tell you about the compound?

When we strike a match there is friction between the match head and the rough edge of the box. This friction makes heat energy. The heat makes the potassium chlorate decompose. It gives off oxygen. The oxygen reacts with the sulphur. The heat from the reaction between sulphur and oxygen is enough to light the carbon on the match head. This gives energy to light the wood of the match stick … and you thought it was simple!

Things to do

1 Copy and complete:
When fuels containing . . . . and hydrogen burn in plenty of oxygen we get carbon dioxide and . . . . formed. This reaction is called complete . . . . However, in a limited supply of . . . . we get the pollution products carbon and toxic carbon . . . . gas released.

2 Find out why carbon monoxide gas is so toxic.

3 What makes a good fuel?
Make a list of the key features you think it should have.

4 Find out about these fuels:

wood ethanol hydrogen

Make a list of the advantages and disadvantages of each one.

Finding the energy

Learn about:
- energy changes in reactions
- using energy changes

Do you enjoy November 5th?
Bonfire night is about chemical reactions!

▶ The chemical reactions in the fireworks are transferring energy.
List all the energy transfers that you can see in the photo.

Most chemical reactions ***give out*** energy.
They are called **exothermic** reactions.
But some reactions ***take in*** energy from the surroundings.
These are called **endothermic** reactions.

Your mouth feels cold when you eat sherbet.
The reaction of sherbet with water is **endothermic**.
The reaction takes in heat energy from your mouth so it cools it down.

Energy in or out?

Try some of these reactions. See what happens to the temperature.

1. Dissolve some citric acid in 50 cm^3 water.
Note the temperature of the solution.
Add crushed limestone to the solution, one spatula measure at a time. Stir and note the temperature after each addition. Add 5 spatula measures in total.
What do you notice?
Is this an exothermic or endothermic reaction?

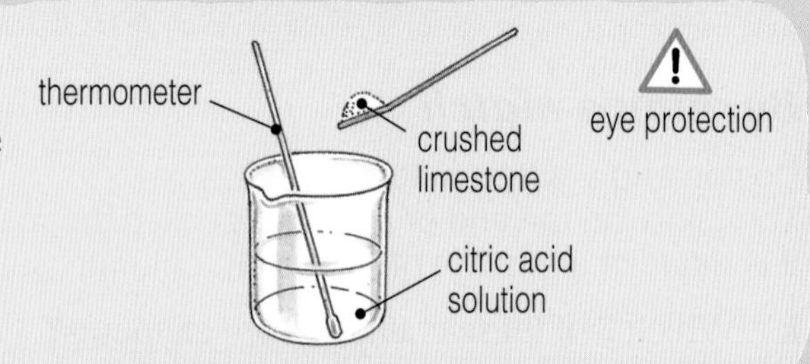

2. Repeat experiment 1 but add sodium bicarbonate (sodium hydrogencarbonate) instead of crushed limestone.
What do you notice?
Is this an exothermic or endothermic reaction?

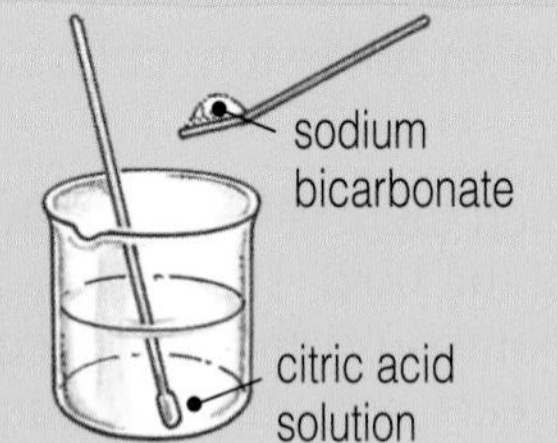

3. Measure 25 cm^3 of dilute sulphuric acid into a beaker.
Note the temperature of the solution.
Add 5 cm^3 of sodium hydroxide solution to the acid. Stir and note the temperature of the solution. Repeat this until 30 cm^3 of sodium hydroxide have been added.
What do you notice?
Is this an exothermic or endothermic reaction?

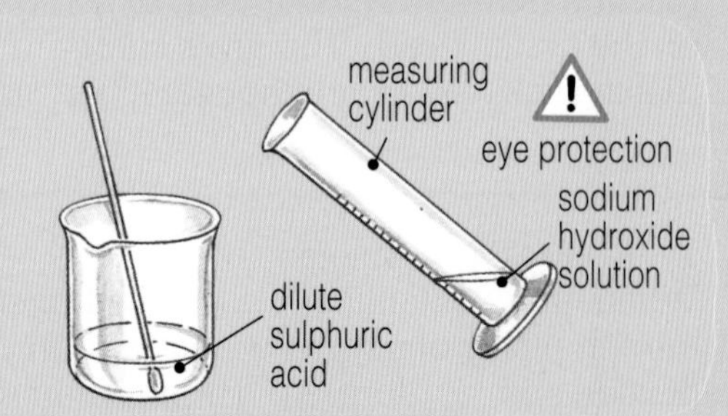

4. Light a Bunsen burner.
Natural gas burns.
What do you notice?
Is this an exothermic or endothermic reaction?

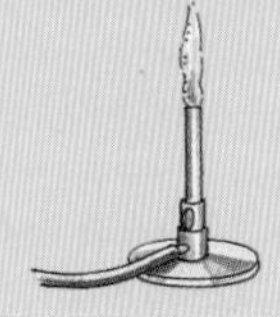

Do ***not*** measure the temperature of the flame. Just observe.

Energy is always transferred when a chemical change takes place.
Sometimes we can observe this
- the temperature may change
- we may see light or hear sound.

Sometimes we can change the stored chemical energy into electrical energy.

Energy from displacement reactions

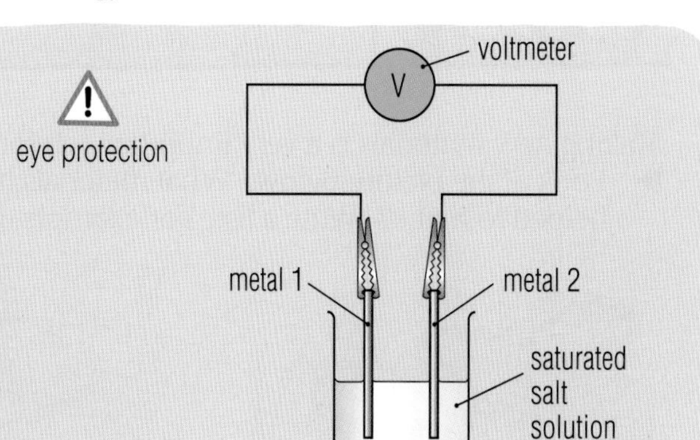

1. Displacement reactions are exothermic.
   Devise an investigation to find if there is a link between the heat produced and the Reactivity Series (see page 74).
   Have your plan checked, then try it out.
2. Set up the apparatus opposite to test pairs of metals.
   Find out if there is a link between the voltage produced and the Reactivity Series.

Think about these energy changes from the chemical reactions below.

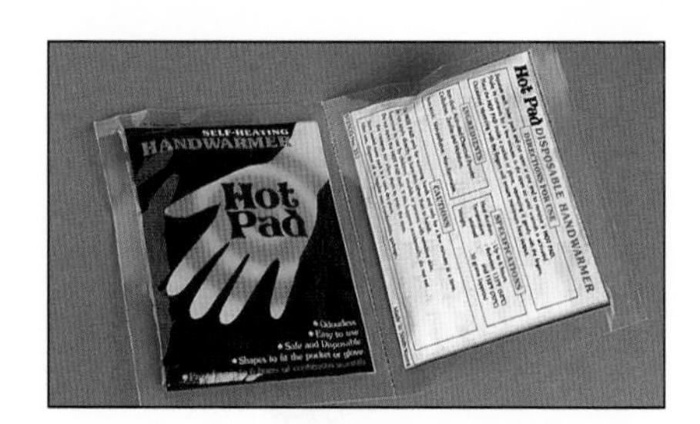

You may have seen these in camping shops.
They keep your hands warm for several hours.

a How do you think they work**?**

Have you ever bought one of these**?**
They are popular at outdoor concerts.

b How do you think they work**?**

c Why is this so useful**?**

But sometimes the energy transfer can be a problem.
Look at the newspaper extract.

d What should you do if you smell a gas leak**?**

Explosion wrecks home!

Workmen who picked their way through the rubble of a house demolished by a gas explosion yesterday found 94-year-old Mrs Ivy Shepherd still standing in the kitchen where she had been eating a cheese sandwich.

The roof and timbers had collapsed around Mrs Shepherd, a widow, but she was unharmed and her only complaint was of the dust in her white hair and a torn apron.

She was taken to hospital in Worthing, West Sussex, for a check-up but was later released.

Things to do

1 Copy and complete:

a) Chemical reactions which give out energy are called . . . .
b) Chemical reactions which take in energy are called . . . .
c) In an . . . . reaction the temperature of the reaction mixture rises.
d) In an . . . . reaction the temperature of the reaction mixture falls.
e) Burning a fuel in air is an . . . . reaction.
f) Neutralisation is an . . . . reaction.
g) Dissolving sherbet in water is an . . . . reaction.

2 Spirit burners can be used to burn liquid fuels safely in the lab.

a) Plan an investigation to compare the amount of energy transferred when different fuels are burnt.

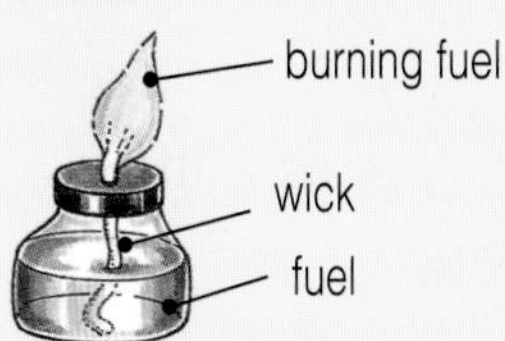

b) Now design an advert for the fuel that gives out most energy. How does the language you use differ in your plan and in your advert?

Scientists – the makers!

Learn about:
- physical changes
- chemical changes

Making new materials is a very important job for a scientist.

▶ Look at the picture below. What materials have scientists helped to make? Write a list. For example, glass for windows.

Do you remember burning magnesium ribbon?
This was an example of making a new substance.

A change which makes a new substance is called a **chemical change**.
A **chemical reaction** must happen.
The new substance is called a **product**.

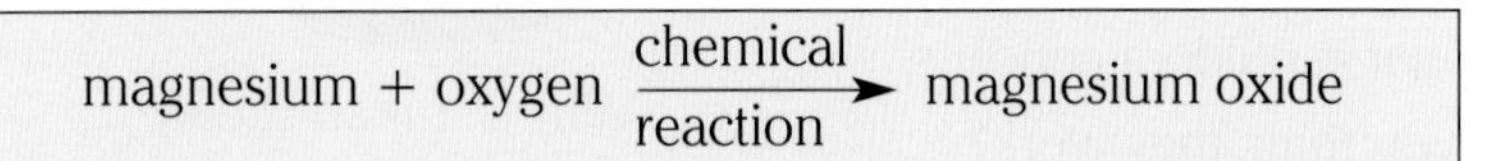

$$\text{magnesium} + \text{oxygen} \xrightarrow[\text{reaction}]{\text{chemical}} \text{magnesium oxide}$$

a What is the product of this reaction?

b How do you know a new substance has been made?

▶ Your teacher may show you 2 changes.
Which one do you think is a ***physical change***?
Which one is a ***chemical change***?
Use the table below to help you.

Physical changes	*Chemical changes*
no new substances made	new substances made
easy to reverse	not easy to reverse

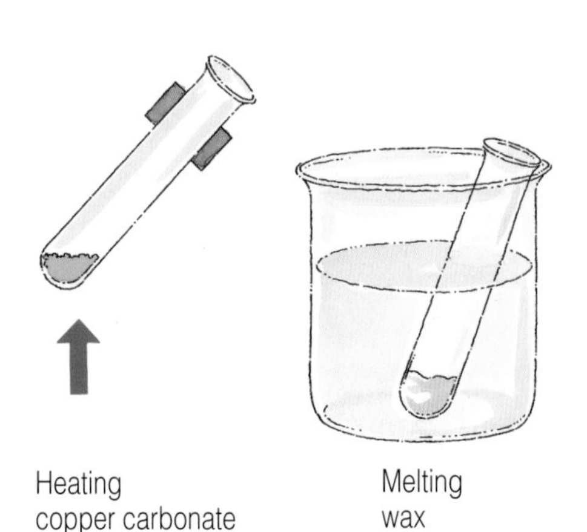

Heating copper carbonate

Melting wax

▶ Which of **c** to **h** are chemical changes?
(Hint: Can you get the materials you started with back again easily? Is it ***reversible***?
or Have new substances been made?)

c Baking a cake.

d Striking a match.

e Making ice cubes.

f Dissolving salt in water.

g Burning some toast.

h Baking clay.

Notice any change?

Here are some tests for you to do. In each one find out what happens when you mix the substances.

For each test, say if you think a new substance is made. Before you start, think about how to record your findings.

- You must wear safety glasses.
- Only use small amounts of the substances. ***Do not use more than the instructions tell you to.***

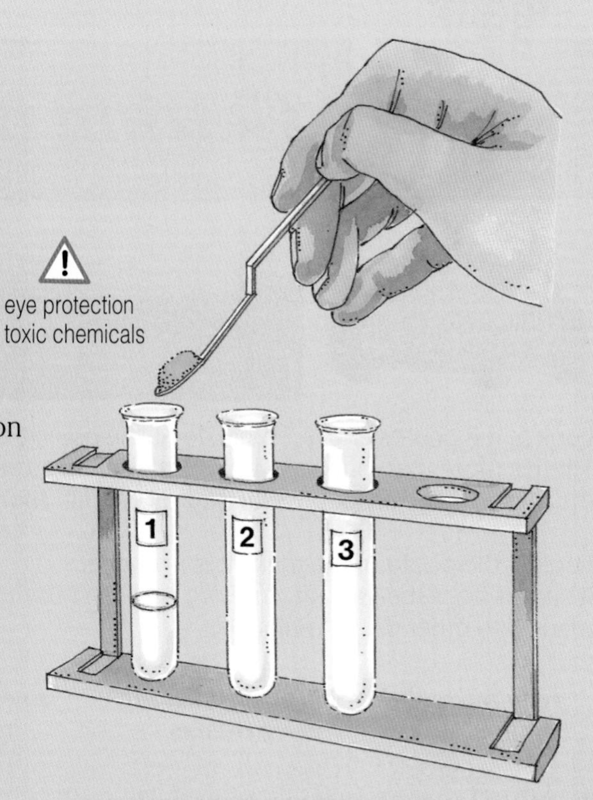

1. 3 cm^3 dilute sulphuric acid + $\frac{1}{2}$ spatula measure of copper carbonate
2. 3 cm^3 dilute sulphuric acid + 3 cm^3 sodium hydroxide solution
3. 3 cm^3 vinegar + $\frac{1}{2}$ spatula measure of bicarbonate of soda
4. 3 cm^3 water + $\frac{1}{2}$ spatula measure of copper oxide
5. 3 cm^3 lead nitrate solution+ 3 cm^3 potassium iodide solution
6. 3 cm^3 dilute sulphuric acid + copper foil
7. 3 cm^3 water + iron nail
8. 3 cm^3 dilute sulphuric acid + 2 cm of magnesium ribbon
9. 3 cm^3 copper sulphate solution + 1 spatula measure of iron filings

Look at your results for all the tests.
How do you know if a new substance has been made**?**

i Make a list of things that can happen when a new substance is made.
These things can tell you that there has been a **chemical reaction**.

Things to do

1 Copy and complete:

a) A change which makes new substances is called a c . . . . change.

b) This change is called a c . . . . r . . . .

c) The new substance is called a p . . . .

d) A p . . . . of the r . . . . between carbon and oxygen is called c . . . . d . . . .

2 Make a list of some chemical changes that happen around your home.
Draw a picture to show one of the changes happening.

3 Do a survey of a car.
Make a list of all the materials used to make the car.
Say whether these are natural or made materials.
Why is each material right for its job?

Material	Natural? or Made?	Why is it used?

4 The Earth's resources, such as sea, crude oil, air, rocks and plants are called ***raw materials***.

Think of a raw material which is made into a useful substance.
Draw a poster to show this change in an interesting way.
What are the uses of the substance made?

Getting the iron

Learn about:
- extracting metals
- breaking down carbonates

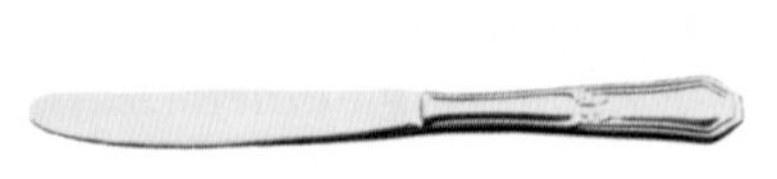

Look at the photos above. What do all the objects have in common? There's more than one answer to this! But if you've read the top of the page, you've probably said they all contain ***iron***. You're right!

Look at the photo of the rock. It is iron **ore**.
It comes from the ground. The ore itself isn't useful. But it can be made into other useful materials.

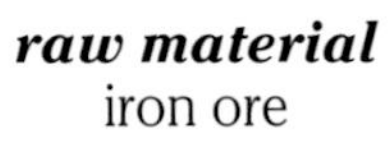
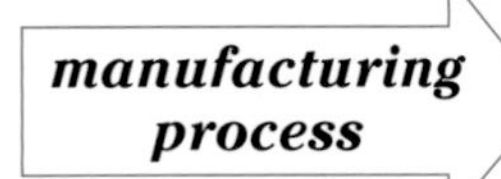

raw material iron ore → ***manufacturing process*** → ***useful products*** iron and steel

▶ Why do you think iron is so useful?

a Make a list of the properties of iron.

b What is steel? Does it have any advantages over iron?

Every day the chemical industry makes many useful products from raw materials. This can involve lots of **chemical reactions**.
Let's look at the reactions needed to make iron.

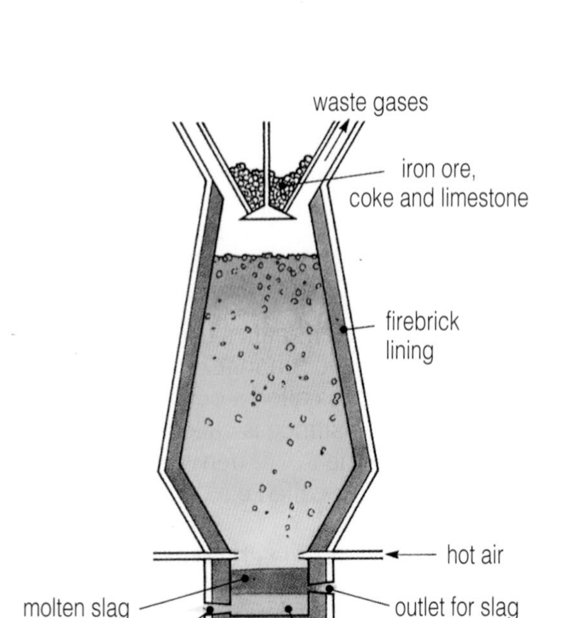

The blast furnace

The most common ore of iron is called haematite. It is iron oxide. This is a compound of iron and oxygen. To get the iron from this ore, we need to remove the oxygen. This is done in a **blast furnace**:

$$\text{iron oxide} \xrightarrow{\text{remove the oxygen}} \text{iron}$$

We say the iron oxide is **reduced**.
Reducing means ***taking the oxygen away***.

In the blast furnace there is some coke, a form of carbon.
This helps to take the oxygen away.
It does this at a temperature of 1200°C.
But there are lots of impurities in the iron ore.
Limestone is used to get rid of these.
In the heat, limestone breaks down. It changes to new substances.

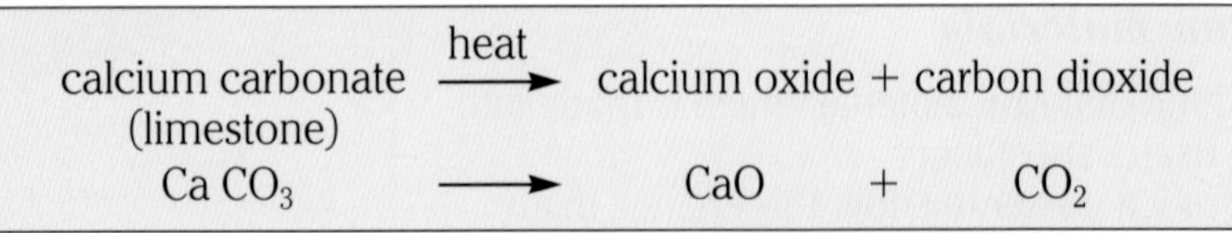

$$\underset{\text{(limestone)}}{\text{calcium carbonate}} \xrightarrow{\text{heat}} \text{calcium oxide} + \text{carbon dioxide}$$

$$CaCO_3 \longrightarrow CaO + CO_2$$

The calcium oxide then reacts with some impure substances in the iron ore. It makes a liquid slag.
It's very hot inside the blast furnace so the iron made is melted (molten).

c Name 3 substances used in the blast furnace.
d Why is the blast furnace lined with fire brick?
e Name one gas which will be a 'waste gas'.
f Why do blast furnaces work every day and night, all year round?
g Why is iron the cheapest of all metals?

A blast furnace

Heating limestone

One of the reactions in the blast furnace involves heating limestone (calcium carbonate). You can see what happens for yourself.

Before you start to heat:

- Ask your teacher to check your apparatus.
- Make sure you know how to stop suck-back (remove the lime water before you stop heating the limestone).
- Think about why you use lime water in the experiment.

suck-back

Heat the limestone gently at first, then more strongly.

h What happens?
i Write a word equation and a symbol equation for this chemical change.

Use this experiment to help you plan an investigation on some other carbonates.

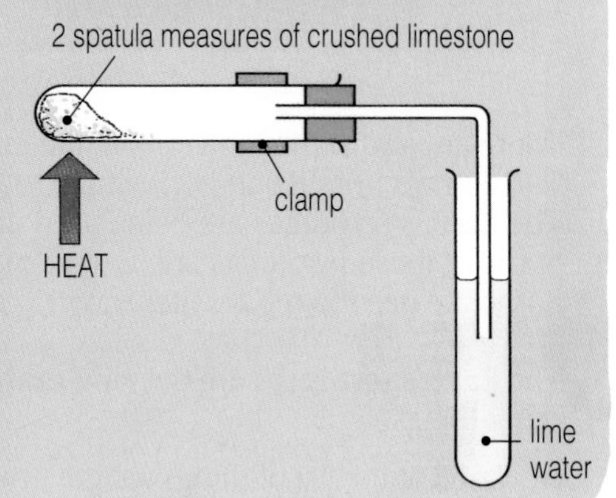

Investigating carbonates

Which carbonate changes the fastest when heated?
Your teacher must check your plan.
If there is time you may be able to carry out the investigation.

j Heat changes all 3 carbonates.
Write word equations and symbol equations for these chemical changes.
k Do you notice any pattern between the reactivity of the metal and how quickly the metal carbonate decomposes?

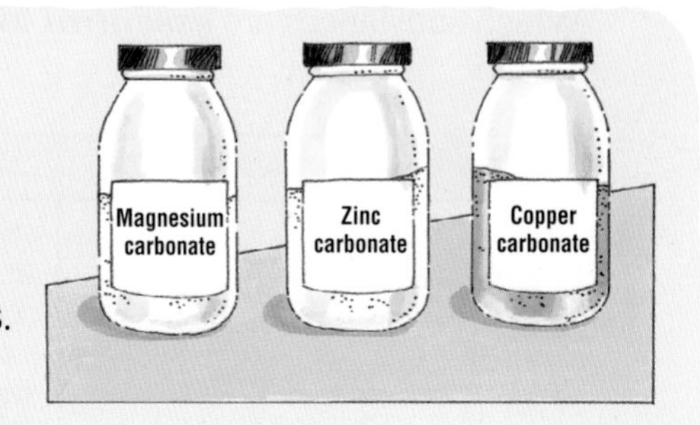

Things to do

1 Copy and complete using the words in the box:

manufacturing	furnace	oxygen
raw	coke	impurities	hot
ore	carbonate	product

A . . . . process changes a . . . . material into a useful . . . . Iron . . . . is made into iron in the blast . . . .
The iron ore is reduced. This means its . . . . is removed. . . . . is a form of carbon which helps to reduce iron ore. Limestone is calcium . . . . It breaks down in the heat. It is used to get rid of . . . . The iron made is melted because it is so . . . . in the furnace.

2 The words in the box are ores.

bauxite, galena, malachite, magnetite

a) Use books to find out which metal comes from each ore.
b) Write down one use of each metal.
c) What things affect the price of any metal?
d) Iron is quite cheap. Name 2 metals that cost more.

3 Limestone is an example of a **sedimentary** rock.
a) Explain how this type of rock forms.
b) Name 2 other types of rock. Write a few lines to explain how each one forms.

Using materials

Learn about:
- making new products

We use up lots of material every day.

▶ Think about what you have done since you got up this morning.
Make a list of all the things you've thrown away.
Could any of these things have been used again?
Do you recycle anything at home or at school?

There are about 7000 million people in the world today.
The world's population is growing.
The Earth's resources are being used up.
Some of these resources are used to make new ***kinds*** of materials.
Some are used to replace items which we throw away.

The Earth's resources are the **raw materials** from which we make other things.

▶ Look at the list of things we get from some raw materials.
In your group discuss whether we need these things.
Which substances are ***essential*** for us to survive? Which are not?

Substance from raw material	*Raw material*
salt	sea
oxygen	air
plastics	coal, crude oil, natural gas
copper	rocks
vegetable oil	living things (plants)

A scientific enquiry

Look at the Task Questions on the opposite page.
Choose 1 of these questions to investigate.
Carry out a scientific enquiry.

Think about:
- where you will get the information from (try to use more than one source)
- how you will present your findings (diagrams, tables, charts or graphs?).

Make sure that you:
- look at all the evidence and information
- make a conclusion (answer the question)
- consider whether your evidence is reliable (are you sure this is the right conclusion?).

Task Question Artificial fertiliser – friend or enemy?

Carry out an enquiry to answer this question.

These are some of the questions you could answer to help your enquiry:

- What is a fertiliser?
- What are the raw materials used to make fertiliser?
- How are artificial fertilisers made?
- Why do we need artificial fertilisers?
- How much artificial fertiliser is used each year in the UK?
- How much does artificial fertiliser cost?
- What are the disadvantages of using artificial fertiliser?

Task Question Recycling – is it worth it?

Carry out an enquiry to answer this question.

These are some of the questions you could answer to help your enquiry:

- What is recycling?
- Which materials are recycled in the UK?
- Why do people choose to recycle?
- What does the recycling process involve?
- How much does it cost to recycle materials?
- In the UK, what percentage of waste is recycled?
- Why isn't more waste recycled?

Things to do

1 Some of the Earth's resources have many uses.
Crude oil is an important resource.
It is a mixture of many substances.
The percentage of each substance is shown in this table:

Name of substance in crude oil	*% of substance in crude oil*
fuel gas	2
petrol	6
naphtha	10
kerosine	13
diesel oil	19
fuel and bitumen	50

a) Draw a bar-chart to show this information.
b) Choose 4 of the substances found in crude oil. Draw pictures to show a use for each of them.

2 Look at the table of raw materials on the opposite page. Which of the 5 could be used to get:
a) sugar?
b) iron?
c) nitrogen?
d) pure water?

3 Some synthetic materials can be helpful or harmful.
Write about the helpful and harmful uses of:
a) detergents
b) medicines
c) explosives
d) pesticides.
Choose one of the materials from a) to d).
Find out how one example of this material was discovered and developed.
Show your findings as a flow diagram.

Investigating mass

Learn about:
- fixed composition of compounds
- conservation of mass

▶ Think about some of the chemical reactions you know.
Try to write word equations for:

a a reaction which can be used to give energy,

b a reaction to make an important new material,

c a reaction which keeps us alive.

How much of each?

Compounds are made when elements join together.
But ***how much*** of each element**?**

Magnesium reacts with oxygen in the air.
But how much of each reacts**?**

magnesium + oxygen ⟶ magnesium oxide

In this experiment you need to know how much magnesium you start with. You then need to know how much magnesium oxide you make.
How do you think you can do this using the apparatus drawn**?**
Discuss this in your group.

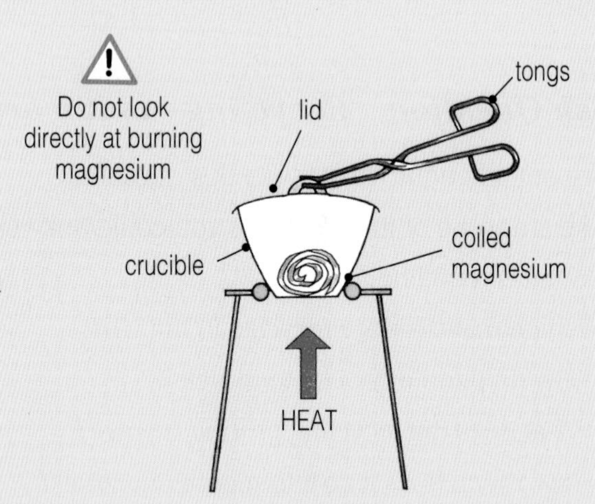

Your teacher will give you an instruction sheet.
Read through all the instructions before you start.
This is a difficult experiment. Your results will tell your teacher how well you have done it!

Each group should have a result.
Copy the class results into a table like this one:

Group	Mass of magnesium at start in grams	Mass of magnesium oxide at end in grams
A		
B		
C		

GROUP	Mass of magnesium at start in grams	Mass of magnesium oxide at end in grams
A	0.6	1.0
B	1.2	2.0
C	1.5	2.5
D		
E		

Draw a graph of the results.

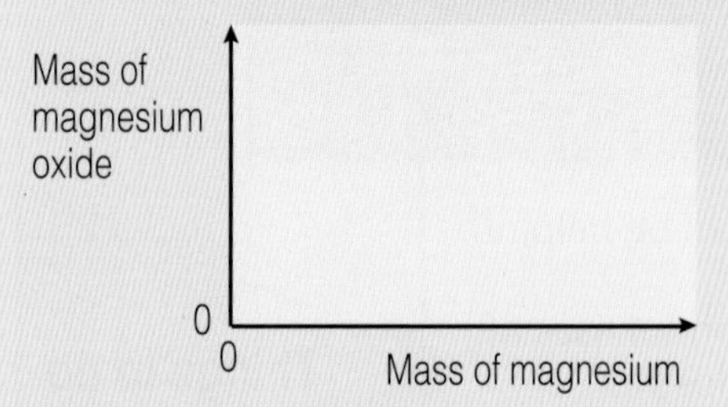

What do you notice? Can you see a pattern in your results**?**

A fixed amount of magnesium always combines with a fixed amount of oxygen.
The compound has **fixed composition**.
The compound always has the formula MgO.

Does the mass change?

Do you think **mass** changes . . . during a physical change**?**
. . . during a chemical change**?**
Predict what your conclusions will be before you do each test.

Test 1 – A physical change

- Put a cube of ice in a beaker.
- Quickly put the beaker on the top-pan balance. Measure its mass.
- Leave the beaker until the cube has half melted. Measure its mass again.
- When the ice has all melted, measure its mass again.

What do you notice about the mass**?**

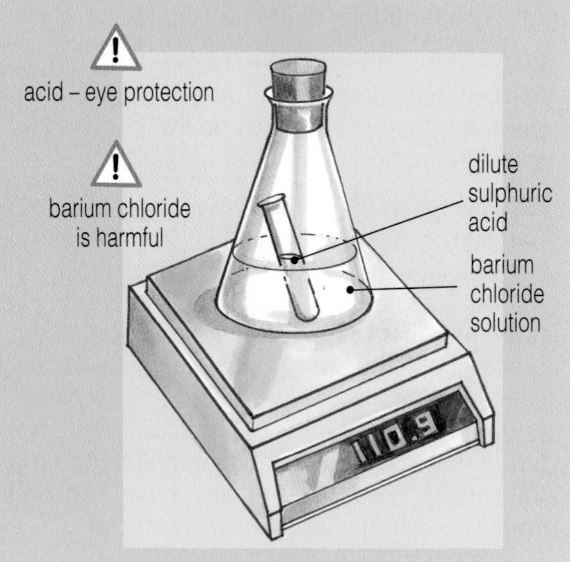

Test 2 – A chemical change

- Half fill a small tube with dilute sulphuric acid. Carefully put it inside a small conical flask.
- Measure out 25 cm^3 of barium chloride solution in a measuring cylinder.
- Carefully use a teat pipette to transfer the solution into the flask. ***Do not drop any solution into the test-tube***.
- Put a stopper on the flask. ***Do not let any acid spill into the solution***. Measure the mass of the apparatus.
- Now tip the flask very gently so the 2 liquids mix. How do you know there is a reaction**?** Measure the mass of the apparatus again.

What do you notice about the mass**?**

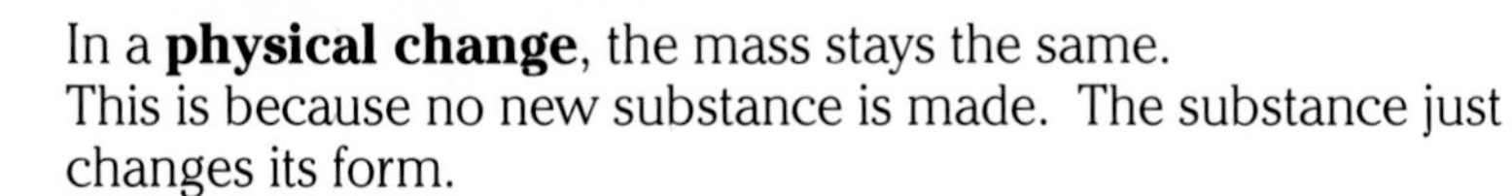

In a **physical change**, the mass stays the same.
This is because no new substance is made. The substance just changes its form.

$$H_2O(s) \longrightarrow H_2O(l)$$
solid ice → liquid water

In a **chemical change**, the ***total*** mass stays the same.
This is because the new substances must be made from the substances already there.
The chemical elements are just combining together in different ways.

$$BaCl_2 + H_2SO_4 \longrightarrow BaSO_4 + 2HCl$$

Joseph Priestley

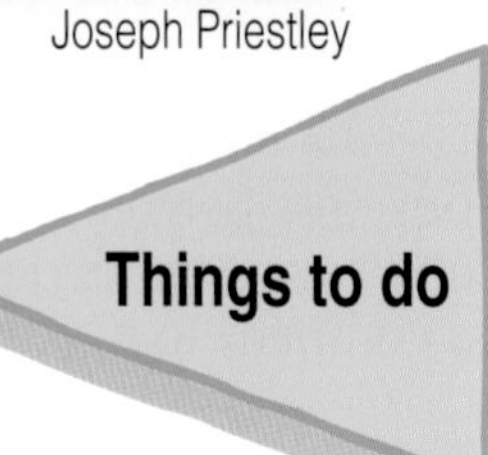

1 Copy and complete:
a) In a . . . . change, a new substance is made.
b) In a . . . . change, no new substance is made.
c) In a chemical change, the total mass . . . .
d) In a physical change, the mass . . . .

2 Melting is a physical change.
a) Describe what happens to the particles as a solid melts.
b) Describe what happens to the particles as a liquid boils.
c) Use your answers to a) and b) to explain why the mass stays the same in a physical change.

3 a) When a candle burns, why does it lose mass?
b) How could you do an experiment with a burning candle to show that the mass of the reactants is the same as the mass of the products in the reaction?
c) In Book 7, you looked at old ideas about burning from the 18th century. Summarise the differences between the phlogiston theory and your ideas about burning now.

4 When hydrogen burns in oxygen, water is made.

$$2H_2 + O_2 \longrightarrow 2H_2O$$

The mass stays the same.
Use the equation to explain why.

Chemical reactions

Learn about:
- the development of a new product
- uses of plastics

The plastics revolution

Leo Baekeland

Have you ever wondered what life would be like without plastics? We can trace the start of the 'plastic age' to a chemist named **Leo Baekeland**. He was born in Belgium in 1863, but worked in America where he made his fortune.

Leo was a chemist with a sharp eye for a business opportunity. His first invention, in the 1890s, was a new type of paper for making photographs. This set Leo up for life when he sold the rights to make the paper for a million dollars.

However, Leo did not settle for a quiet life in his mansion in New York State. In a converted barn on his estate, he set about his next project. At that time, ten-pin bowling was becoming very popular in the USA. The wooden floors of the bowling alleys were varnished with a substance called shellac. This was imported from Asia, where beetles left a sticky liquid on trees. The liquid was collected, heated and filtered to make the varnish. Leo tried to make a new varnish that could be manufactured from chemicals.

Leo made his first million dollars from photographic paper

He investigated chemicals you could get from coal-tar and wood alcohol. Thirty years earlier a German chemist had discovered a thick, gooey substance as he tried out reactions to make new dyes. To Adolf von Baeyer this was just a nuisance. But Leo could see its potential for use as his new varnish. Not only would it help the budding sport of ten-pin bowling, but it would signal the start of the plastics industry.

When Leo heated up the gooey liquid it turned even thicker. If he did the reaction under pressure it made a hard, translucent solid. He could mould this into any shape he liked. By 1907, he had made the first synthetic plastic! It had taken him 3 years and thousands of failed experiments, but his perseverance paid off in the end.

Leo the chemist quickly became Leo the business-person. He patented his new plastic and gave it the trade name 'Bakelite'. He formed a company to make and sell the plastic.

Ten-pin bowling spurred Leo on to make the first synthetic plastic

A new customer would help Leo make another fortune.
The electricity industry was just starting to grow.
Plans to get electricity into homes were being drawn up, but there was a problem. The early insulation material used was shellac.
Leo could make moulded electrical insulators from his new plastic.
By doing this he played a big part in changing life in the last century.

People were quickly using the new plastic for everything from buttons to telephones. You might have seen old-fashioned radio or TV sets made from a hard brown plastic. This was Bakelite. But Leo could make his plastic in a variety of colours. It even became trendy to wear Bakelite jewellery. Plastics had arrived in our lives!

Bakelite was used in the early radios and TVs

Look at the variety of plastics we have developed now and some of their uses:

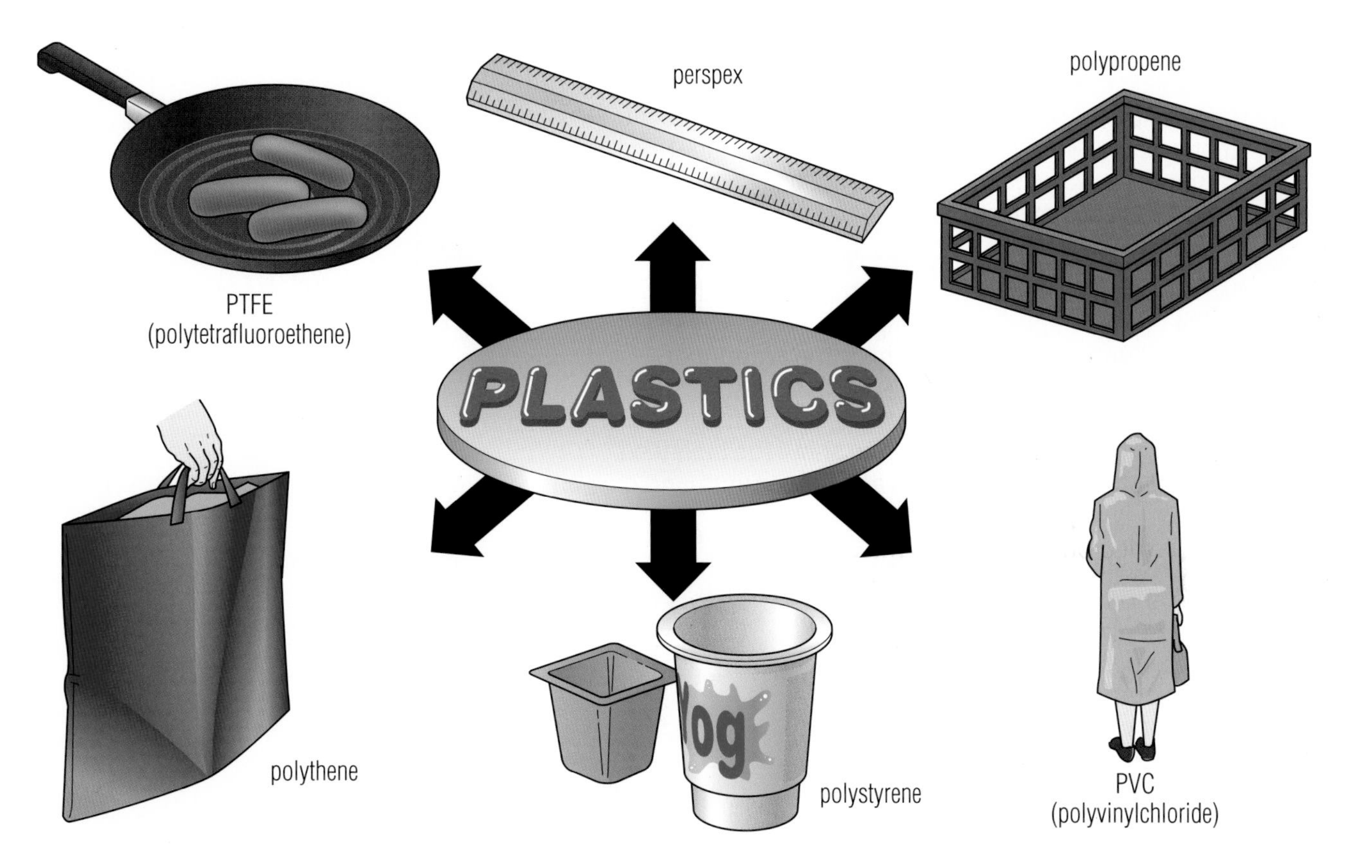

Questions

1 How was Leo Baekeland's chemical research influenced by business opportunities?

2 What do you think were the characteristics of Leo Baekeland that made him a wealthy man?

3 Make a list of all the plastic things you have used so far today.
How would your life today be different without plastics?

4 Most plastics nowadays are made from products we get from crude oil.
How do you think the plastics industry will develop in the next 50 years?

5 Do some research to find out about either:
a) The problems of plastics waste and how we might solve them, or
b) The life and work of Wallace Carothers.

Questions

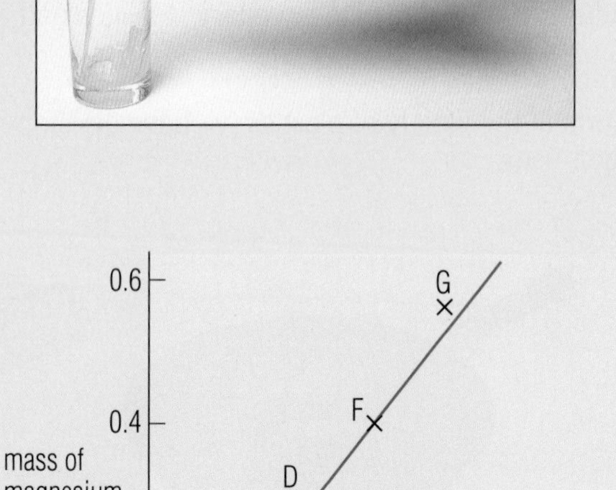

1 Which of these are physical changes?
a) Adding water to orange squash.
b) Burning petrol in a car engine.
c) Getting salt from sea water.
d) Making detergent from crude oil.

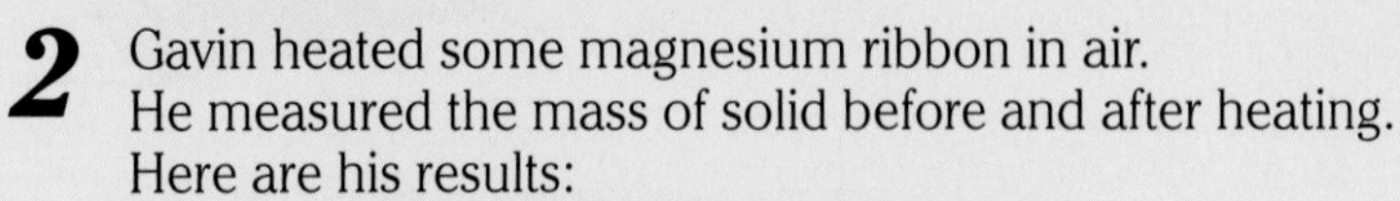

2 Gavin heated some magnesium ribbon in air.
He measured the mass of solid before and after heating.
Here are his results:
Mass of solid before heating = 0.24 g
Mass of solid after heating = 0.40 g
a) Copy and complete the word equation for the reaction:
magnesium + ... ⟶ magnesium oxide
b) Name the solid product of the reaction.
c) Try to explain Gavin's results.

Gavin made a graph of his class results:
d) Which group had a strange (anomalous) result?
What do you think went wrong?
e) Which 2 groups started with the same amount of magnesium?
f) The chemical formula of oxygen is O_2 and magnesium oxide is MgO. Draw a diagram to show the balanced equation for the reaction between magnesium and oxygen.

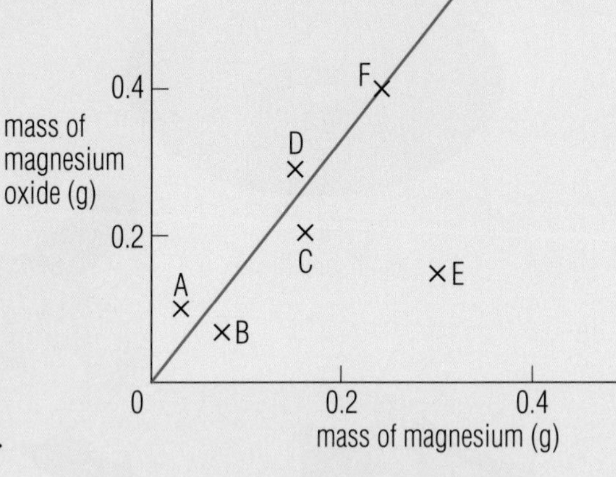

3 Chemical reactions don't always happen.
a) Think about the reactivity series. Predict whether a reaction will take place in each case.
i) magnesium + magnesium sulphate
ii) zinc sulphate + copper
iii) copper sulphate + zinc
iv) copper sulphate + magnesium
b) Write a word equation for the reactions you predict will happen.
What do we call this type of reaction?
c) Predict which reaction from part a) will give out most energy.
Explain your answer.
d) What do we call reactions that give out energy?

4 Handwarmers are okay, but . . . .
Invent some self-warming soup for winter walks.
a) Design a can of soup which can warm itself when opened.
b) Make an advertisement for your invention.

5 Can scientists help to solve some of the world's problems?
Imagine ***you*** could make some new materials to solve these problems.
What would you want your new materials to do?
Make a list of your ideas.

Energy and electricity

Electricity is important to all of us.
Our lives would be very different without it.

We use electricity to transfer energy from one place to another.
We use it for lighting and for heating.

In this unit we also see how we use energy to generate electricity.

Go with energy

Learn about:
- energy transfers
- using electrical energy

a Name 4 things in your home that use electricity.

b Why do we need to eat food?

c Make a list of 5 things that you have done today. Put them in order, starting with the one that you think needs most energy.

If you climb to the top of a ladder, you have more gravitational **potential** energy.
An object which is moving has **kinetic** energy.

d Complete this sentence:
When a skier is moving downhill, ... energy is transferred to . . . . energy.

▶ Look at the diagram. It shows an electric motor lifting up a weight:

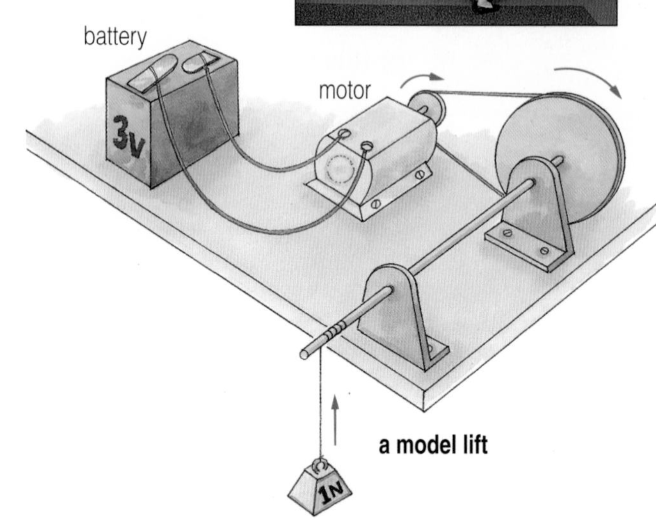

a model lift

e What kind of energy does the moving wheel have?

f Where does this energy come from?

g Copy and complete its **Energy Diagram**:

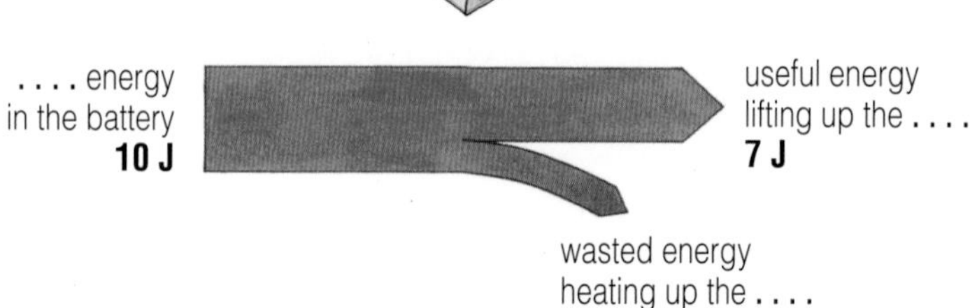

h From the diagram, what can you say about the amount of energy (in joules) ***before*** the transfer and ***after*** the transfer?

i How much of the energy after the transfer is useful? What has happened to the rest?

Only 7 out of 10 joules have done something useful.

This is what usually happens in energy transfers. Although there is the same amount of energy afterwards, not all of it is useful.

▶ Now look at this diagram:

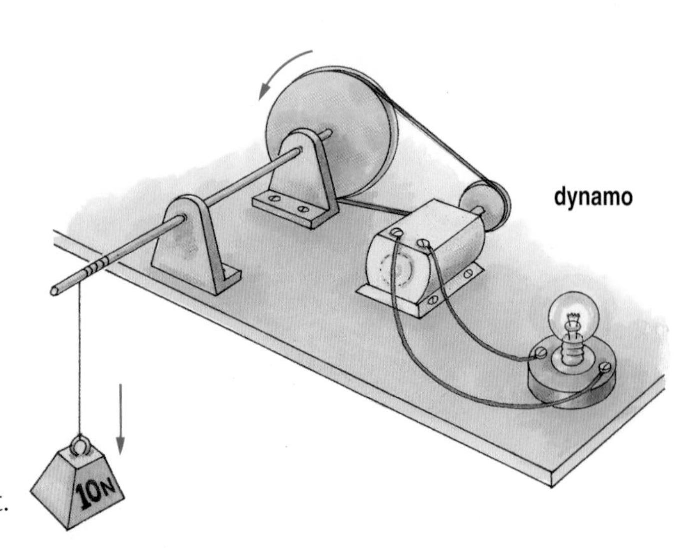

j What happens to the lamp? Why?

k Suppose the weight starts with 100 J of potential energy, and then 20 joules appear as light energy shining from the lamp. What has happened to the other 80 joules?

l Draw an Energy Diagram for this, and then label it.

m Name the forms of energy in each part of the equipment.

Energy transfers

Your teacher will show you some of these.
Observe them carefully, and think about the energy transfers.
For each one, draw an Energy Diagram and label it.

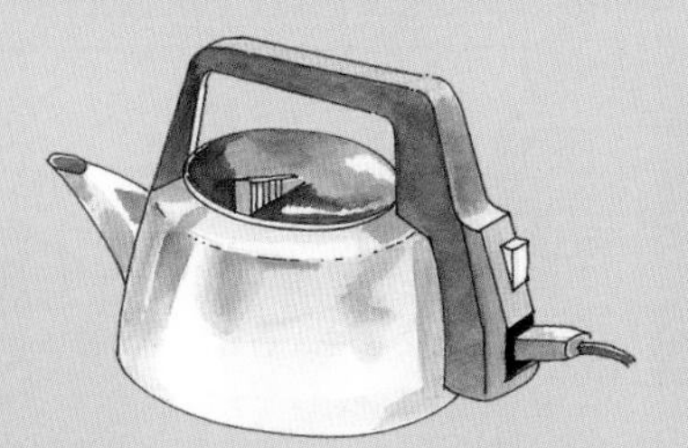

n an electric kettle

o a clockwork toy

p a solar-powered calculator

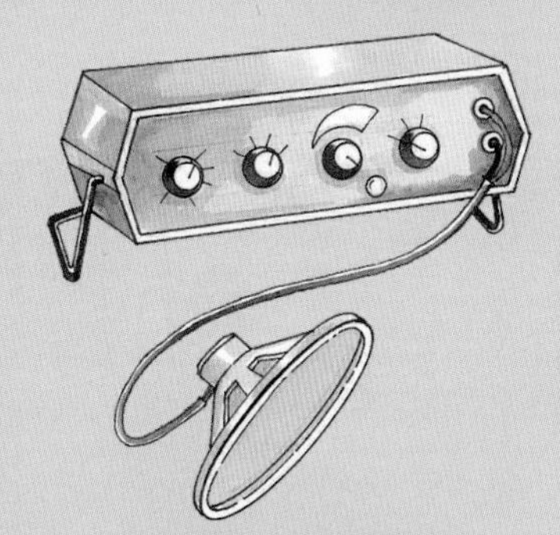

q a signal-generator and loudspeaker

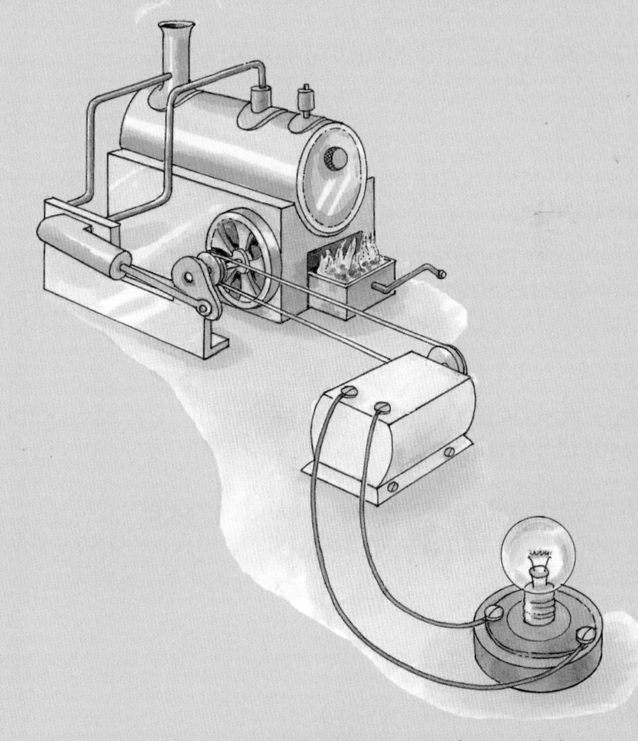

s a steam engine and dynamo

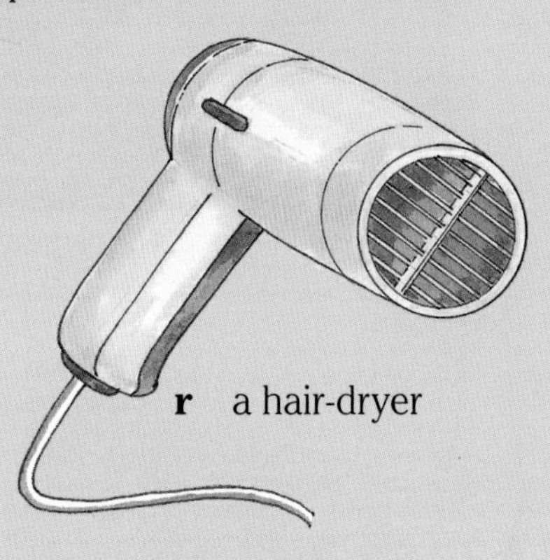

r a hair-dryer

Things to do

1 Copy and complete:

a) Energy is measured in . . . .
b) Lifting a stone gives it . . . . energy.
c) A moving object has . . . . energy.
d) In any energy transfer, the total amount of . . . . before the transfer is always . . . . to the total amount of . . . . afterwards.
e) After the transfer, not all of the . . . . is useful.

2 Describe the energy transfers involved in:

a) a torch,
b) a TV set,
c) playing a guitar.

3 Give 3 examples of ways in which energy can be stored.

4 Petrol (chemical energy) is put in a car. Give examples from different parts of the car of the forms that energy can be changed to.

5 Are some kettles cheaper to run than others? Plan an investigation to compare some kettles.
How will you make it a fair test?

6 Why do you think it is impossible for anyone to build a perpetual-motion machine?

Investigating batteries

Learn about:
- measuring voltage
- making a battery

Batteries are a useful way to make electricity. But expensive!

They have chemicals inside, to store energy.
When you use them in a circuit, the chemical energy is transferred to electrical energy.

▶ Make a list of all the things you can think of that use batteries.

▶ Draw a circuit diagram for a torch.

▶ Draw an Energy Transfer Diagram for a torch.

A battery or 'cell' makes an electric current until the chemicals in it are used up.

Some batteries are ***re-chargeable***. They can be re-charged with electricity so that they work again.
A car battery is re-chargeable.

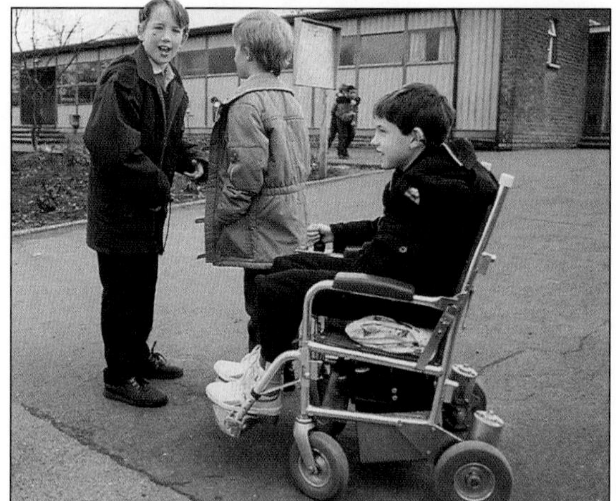
Using a rechargeable battery

A battery pushes electrons round the circuit. The size of the push is measured in **volts**.

Batteries push with just a few volts and are safe.
Mains electricity pushes with 230 volts (230 V) and so it is very dangerous!

Using a voltmeter

Connect a **voltmeter** to a battery. Take care to connect the red (+) terminal of the voltmeter to the + (the button) of the battery. What do you see**?**

The voltmeter measures how hard the battery pushes the electrons.
What is the voltage of a simple battery (a 'dry cell')**?**

What do you think will happen if two cells in series push the ***same*** way**?** Try it. What do you find**?**

What do you think will happen if two cells in series push in ***opposite*** directions**?** Try it. What happens**?**

What is the voltage of each battery in the photograph at the top of this page**?**

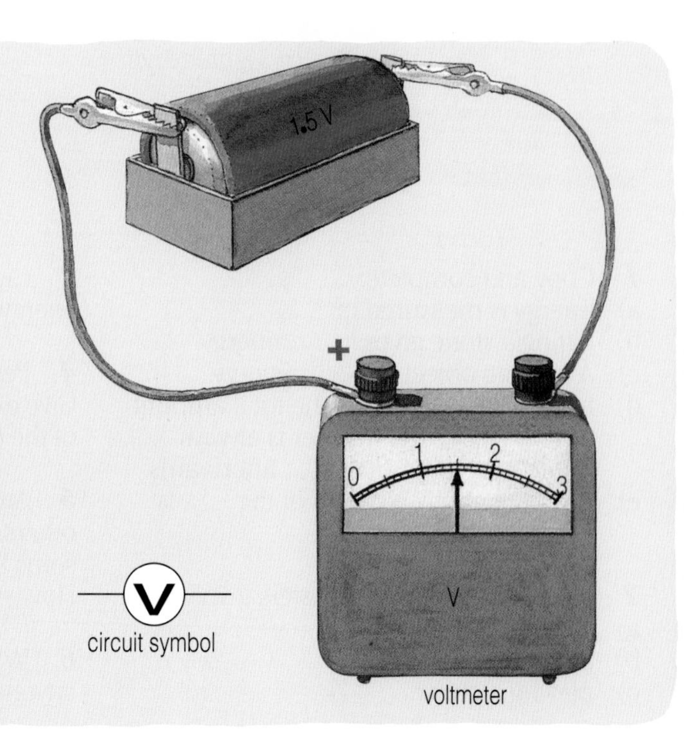

Fruity batteries

You can make a simple battery (a cell) by pushing 2 different metals into a fruit:

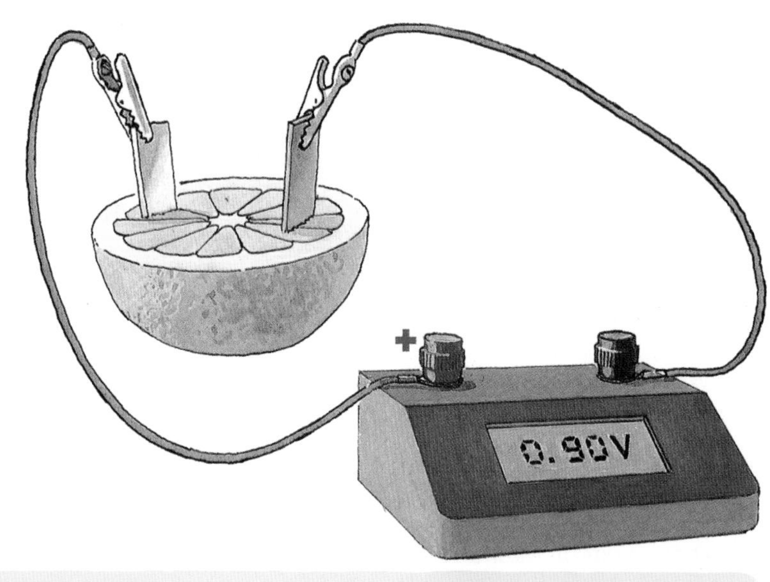

You are going to investigate how to get the highest voltage.

▶ Choose ***one*** of the investigations.

Then plan it:

- Decide what you are going to change each time.
- Decide what you are going to measure each time.
- Decide what you must keep the same each time, to make it a fair test.
- Decide how to record your results.

Show your plan to your teacher and then do it.

Investigation 1

Which 2 metals give the highest voltage**?**

Investigation 2

Which fruit gives the highest voltage**?**

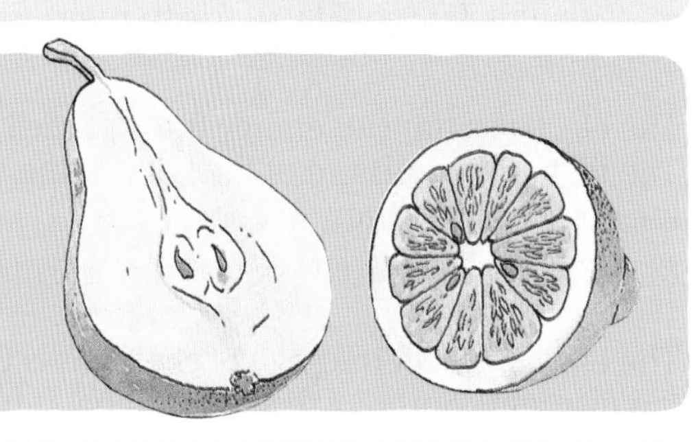

Investigation 3

Which fruit and metals give the highest voltage**?**

Afterwards:

- Write a report on what you did and what you found out.
- Draw a bar-chart of your results.
- Which gave the highest voltage**?**

Things to do

1 Copy and complete:

a) A battery pushes . . . . round a circuit. The size of the push is measured in . . . . , by using a . . . .

b) Mains voltage is . . . . volts and so is very . . . .

c) If two 3 volt batteries push together in the same direction, then the total voltage is . . . . volts.

2 Draw a circuit diagram of a battery connected to a voltmeter, with a switch in the circuit.

3 A 9 V battery is made from small 1.5 V cells. Draw a diagram to explain this.

4 High voltages can transfer large amounts of energy. Where can you find high voltages? What are the hazards of high voltages?

In series

Learn about:
- energy from cells
- series circuits

▶ The diagram shows a circuit with 2 bulbs in **series**:

a What does each symbol stand for?

b What can you say about the current through **A** and the current through **B?**

c What happens if bulb **B** breaks? Why?

d Draw a circuit diagram of 3 bulbs in series.

Bulbs in series

Energy from a battery

A cell pushes electrons (−) round the circuit:

The flow of electrons is called a **current**. The size of the current is measured in amperes (amps, A).

e What meter would you use to measure the current?

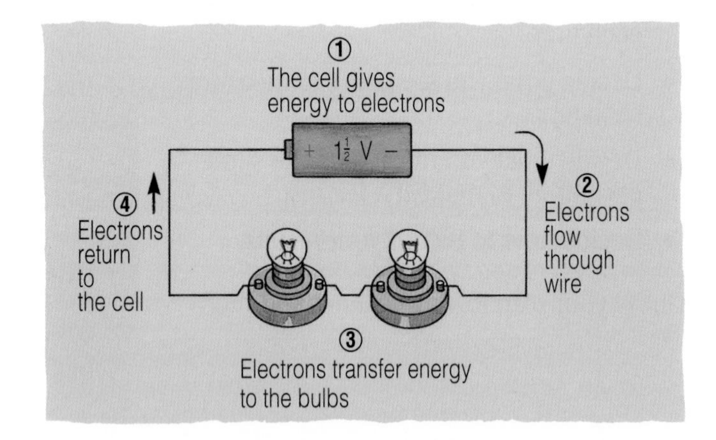

The electrons transfer energy from the cell to the bulbs. The higher the **voltage** of the battery, the more energy the electrons can transfer:

f What meter would you use to measure the voltage?

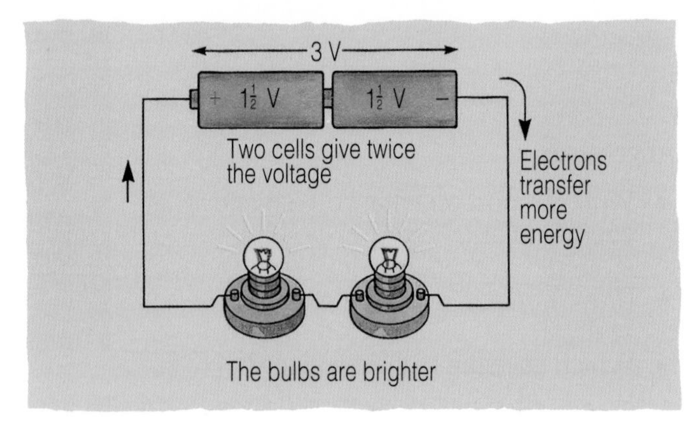

The energy transferred by the electrons is used to heat up the filament. Some of this heat energy is transformed to light energy, which is then transferred by radiation.

g Draw an Energy Transfer Diagram for this circuit.

A conductor lets electrons pass through it easily. It has a low **resistance**.
An insulator has a high resistance. The electrons cannot move through the insulator, so there is no current.

h Write down the names of 3 conductors and 3 insulators.

i 'Mains' voltage is 230 V. Why is this dangerous?

▶ Here are some of the symbols used in circuit diagrams.

j Copy the symbols and label each one.

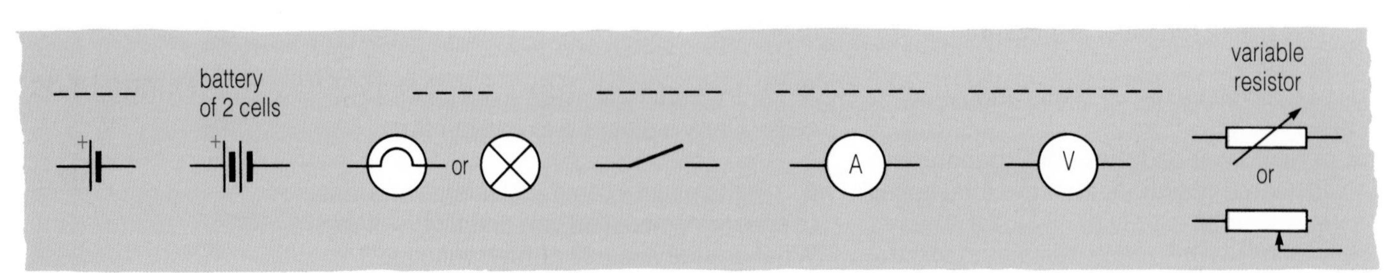

Reading ammeters

The diagrams show two ammeters:

What are the readings **k**, **l**, **m**, **n**, **o**, and **p?**

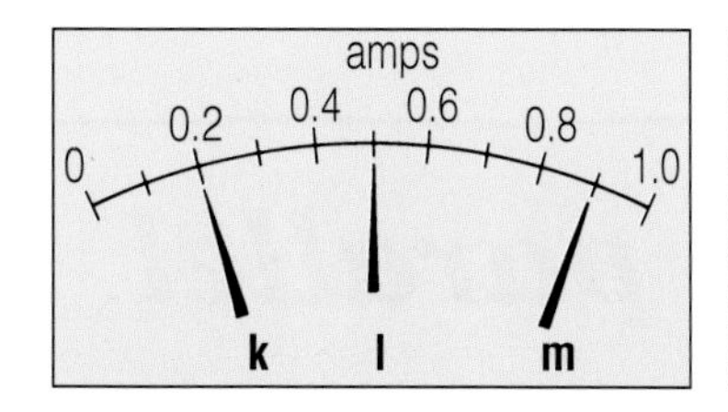

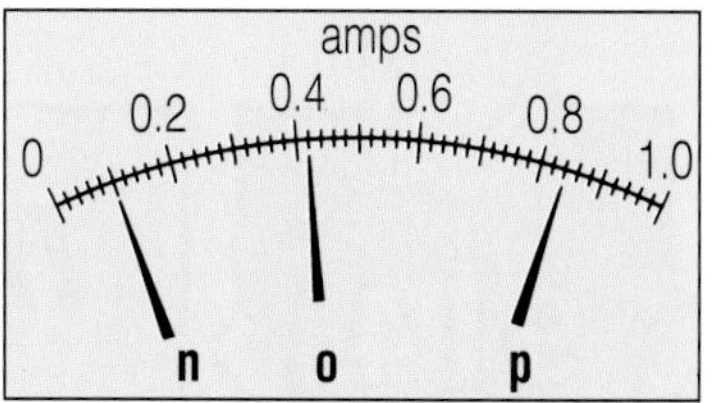

Measuring currents

Here is a series circuit:

- Draw a circuit diagram for it.
- Then connect up the circuit. Make sure the + terminal of the ammeter is connected to the + terminal of the battery.
- Adjust the variable resistor to make the bulb as bright as possible. What is the reading on the ammeter**?**
- Disconnect the ammeter and then re-connect it in position 2. What is the reading now**?** What is the reading on the ammeter if it is in position 3**?** Write a sentence to describe what you found.
- Now connect a second bulb in ***series***. What happens to the brightness of the first bulb**?** What happens to the current through the ammeter**?** Explain why you think this happens, using the words: **current electrons resistance**
- Predict what would happen if you added a third bulb in series. Try it if you can.
- With just one bulb, use the variable resistor to reduce the current until the bulb is not quite glowing. What is the reading on your ammeter now**?**

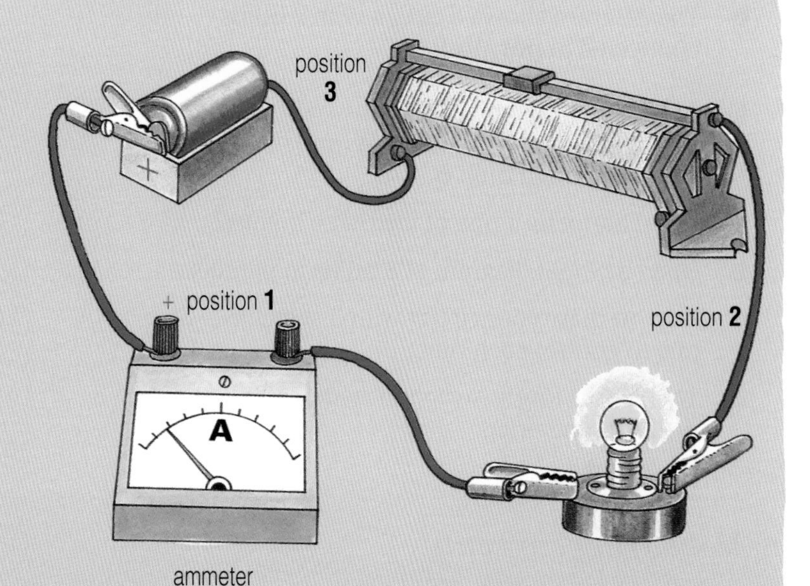

Position of ammeter	*Current (amps)*
1	
2	
3	

A digital ammeter

Things to do

1 Copy and complete:
In a **series** circuit,
a) The current is the . . . . through each part of the circuit. It is not used up.
b) If you add extra bulbs, the current is . . . . and the bulbs are . . . . bright.
c) If you add extra cells, the electrons have . . . . energy and bulbs shine . . . . brightly.
d) An ammeter measures the . . . . in a circuit, in . . . . or A.
e) A battery pushes . . . . round a circuit. The size of the push is measured in . . . . , by using a . . . .
f) A good conductor has a . . . . resistance.

2 An ammeter is connected in series with a battery, a switch and a bulb.
a) Draw the circuit diagram.
b) If the ammeter reads 0.8 A, how much current passes through the bulb?
c) A second bulb is connected in series. Draw the circuit diagram.
d) The ammeter reading is now 0.4 A. How much current passes through each bulb now?
e) Explain why this current is less than before.

3 The ampere and the volt are named after André Ampère and Alessandro Volta. Find out more about these people and what they did.

In parallel

Learn about:
- parallel circuits
- calculating currents
- calculating voltages

▶ The diagram shows a circuit with 2 bulbs connected in **parallel**:

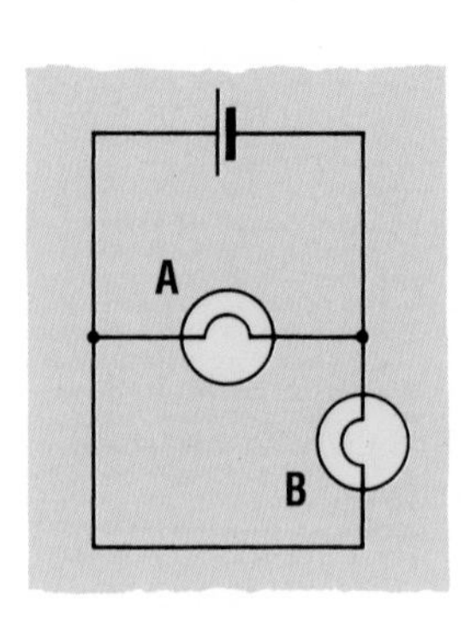

Bulbs in parallel

When the electrons travel from the battery, ***some*** of them go through bulb **A**, and the rest of them go through bulb **B**.

a What happens if one of the bulbs breaks**?**

b Draw a circuit diagram of 3 bulbs in parallel.

c Now re-draw your circuit with 3 switches, one to control each bulb.

d Then add a fourth switch to switch off all the bulbs together.

Measuring currents

- Connect up this circuit, with the **ammeter** in position 1:
  Take care to connect the ammeter correctly.
- What is the reading on your ammeter**?**
- Then connect the ammeter in position 2.
  What is the current through bulb **A?**
- Then find the current in position 3.
- What do you notice about your results**?**
  Explain this, using these words:
  current **electrons** **resistance**

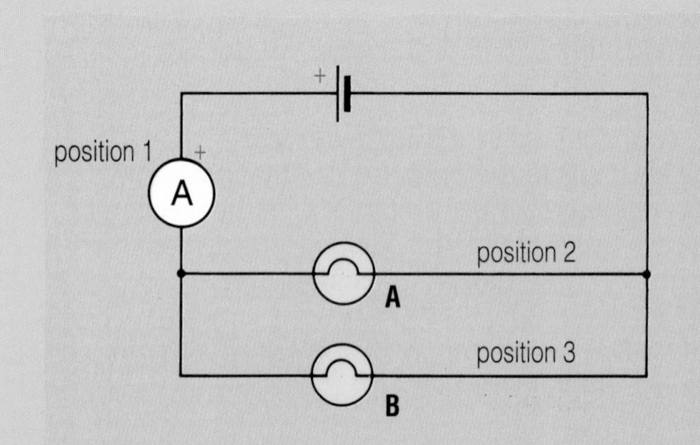

Notice the ammeter is always in series in the circuit.

Calculating currents

Look carefully at this circuit:

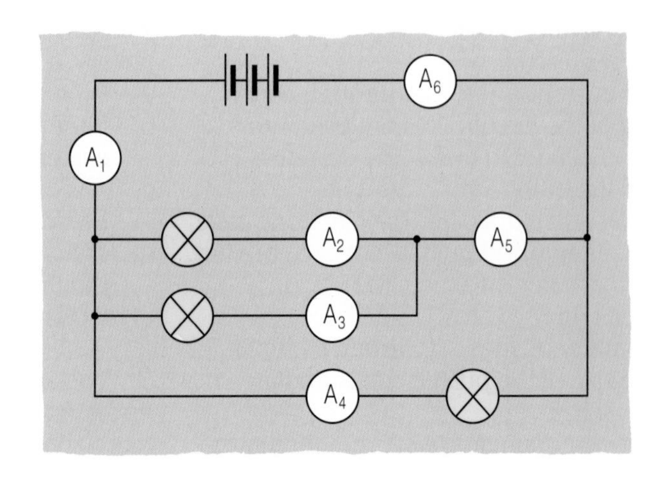

Suppose you are told the readings on 3 of the ammeters are:
$A_2 = 1$ A $A_3 = 1$ A $A_4 = 1.5$ A

e What is the reading on ammeter A_5**?**

f What is the reading on ammeter A_6**?**

g What is the reading on ammeter A_1**?**

h If more cells are put in the circuit to increase the voltage, what would you expect to happen to each of the ammeter readings**?**

Measuring voltages – 1

Connect up this circuit with a cell and 2 bulbs in parallel:

A **voltmeter** is connected ***in parallel*** with the bulbs.

Use your voltmeter to measure:

- the voltage across bulb **1** (V_1),
- the voltage across bulb **2** (V_2),
- the voltage across the cell (V_3).

What do you find?

Can you explain this?

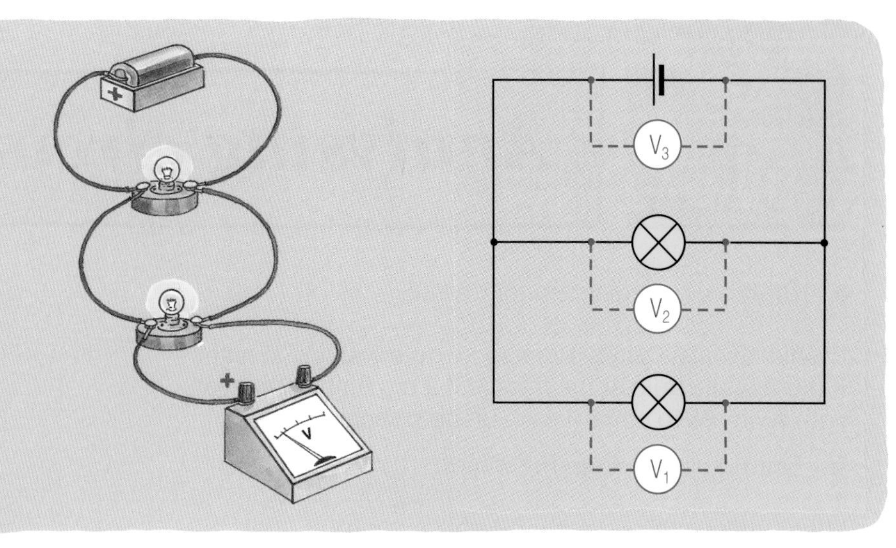

Measuring voltages – 2

Now connect 2 bulbs in series with a cell as shown:

Use your voltmeter to measure:

- the voltage across bulb **1** (V_1),
- the voltage across bulb **2** (V_2),
- the voltage across the cell (V_3).

What do you notice about your results?

Try a third bulb in series. What do you find now?

Notice that the bulbs are in series but the voltmeter is ***always in parallel*** with a component.

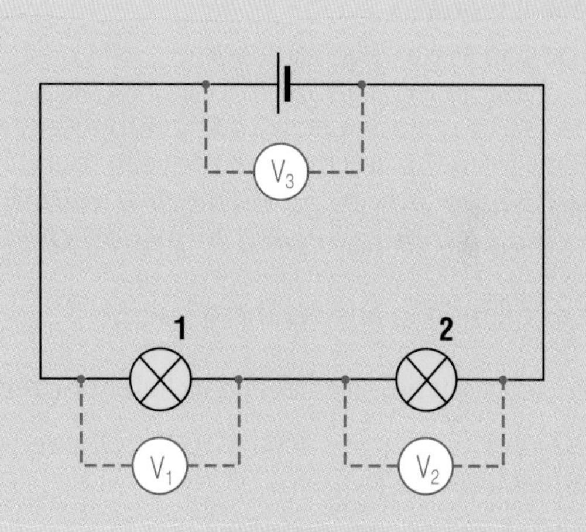

Calculating voltages

In this circuit, a voltmeter connected across the cell reads 3 V. A voltmeter placed across bulb **A** reads 2 V.

i What is the voltage across bulb **B**?

j Which bulb do you think gets more energy? Why?

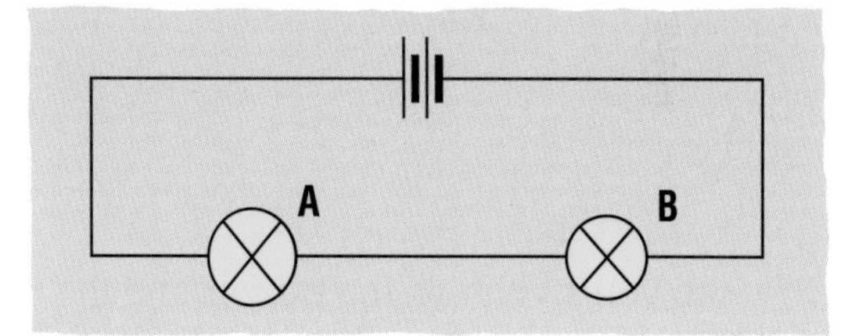

Things to do

1 Copy and complete:

a) An ammeter measures the . . . . passing through a component, so it is always put in . . . . with the component.

b) A voltmeter measures the . . . . across a component, so it is always put in . . . . , across the component.

2 Here is a circuit with 3 ammeters:

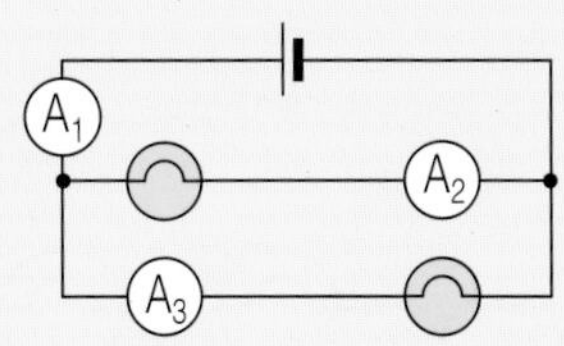

If ammeter A_1 shows 0.5 A and A_2 shows 0.3 A, what is the reading on A_3?

3 Draw a circuit diagram to show how two 6 volt bulbs can be lit brightly from two 3 volt cells.

4 Here is a circuit with 4 voltmeters:

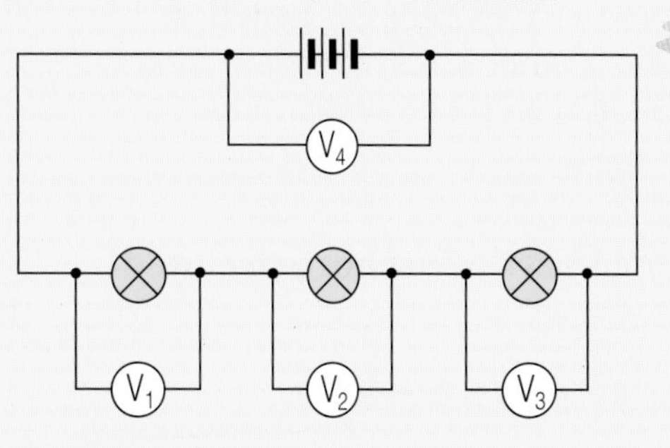

If voltmeters V_1, V_2, V_3 all show 2 V, what is the reading on V_4?

Analysing circuits

Learn about:
- more complex circuits
- solving problems

a Draw a circuit diagram for a torch.

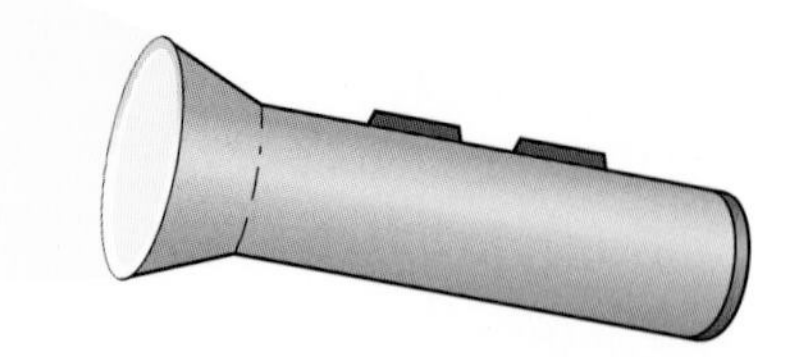

Rashid invents a safety torch for people walking at night. It has a white bulb at the front and a red bulb at the rear, with switches to control each of them separately.

b Draw a circuit diagram for him.

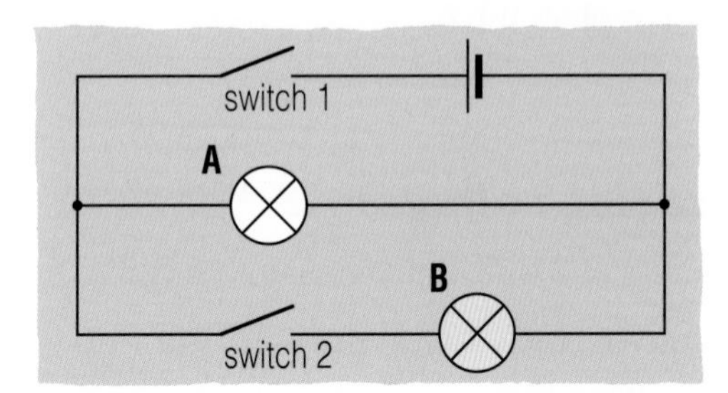

Luke suggests this circuit for Rashid's torch:

c Will it work?

d Can you see a disadvantage?

To analyse the circuit, use your finger to follow the path of the electrons from the cell through the bulb A and back to the cell. ***If your finger has to go through a switch, then this switch is needed to put on the light.***

Use this method to answer these questions:

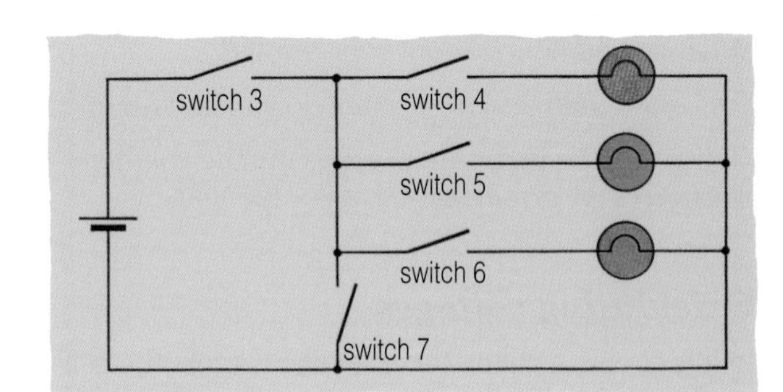

e Which switches are needed to light the green lamp?

f Which switches are needed to light the red and blue lamps together?

g What happens if both switches 3 and 7 are closed?

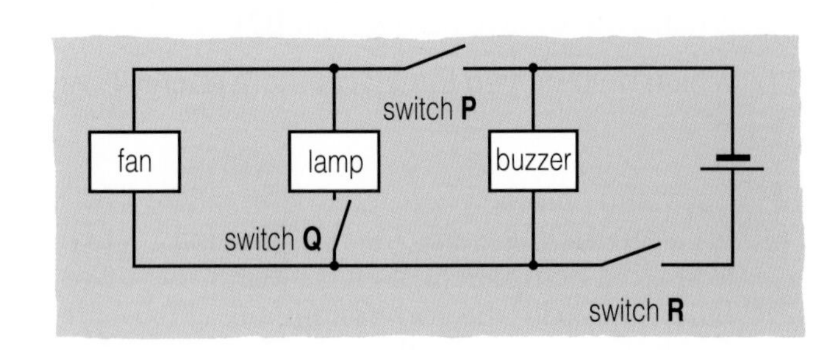

h How can you switch on the buzzer and the lamp?

i What happens if only switches **P** and **R** are closed?

j Re-draw the circuit so that the fan, lamp and buzzer each have their own switch.

k Draw a circuit with a battery of 2 cells, a switch, an ammeter, 2 bulbs, and a variable resistor, all in series. Add a voltmeter in parallel with one of the bulbs.

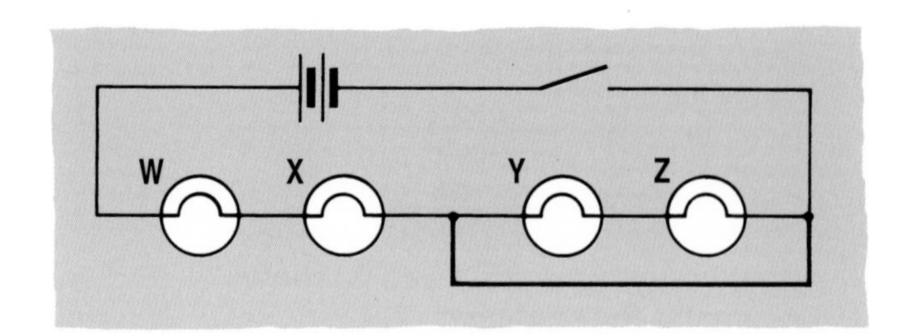

In this circuit, 2 of the lamps have been **short-circuited** by a thick wire:

l What happens when the switch is closed?

m Re-draw the circuit so only lamp **Z** is short-circuited.

Two-way switches

The diagram shows a switch that can be in position A ***or*** in position B.

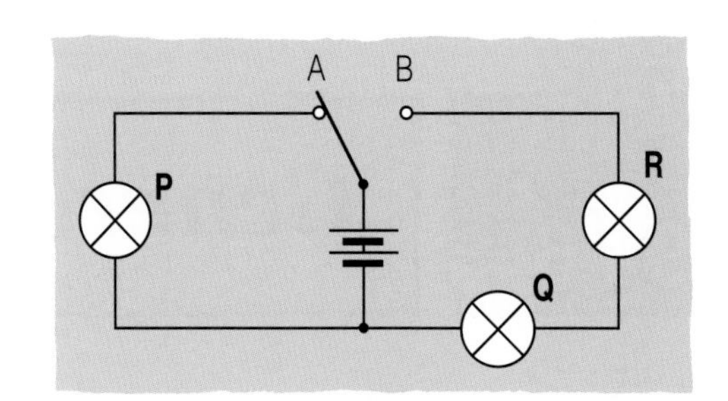

n What happens if the switch is in position A**?**

o What happens if the switch is in position B**?**

The diagram shows a mains circuit in a house, with two 2-way switches:
Wire C can be connected to either A or B.
This circuit is often used for the lighting on a staircase.

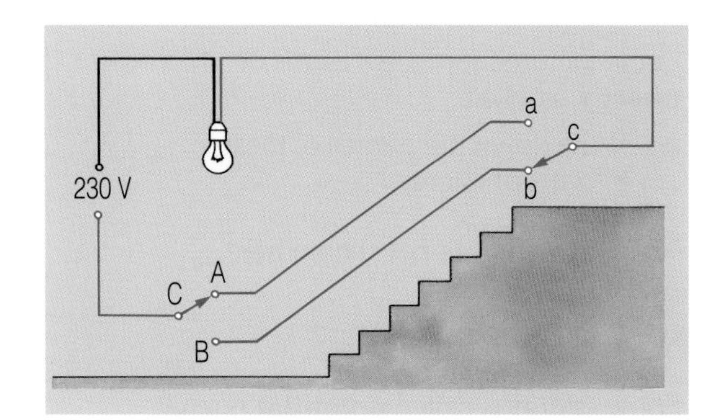

p In the diagram, is the lamp on or off**?**

q Describe how the circuit works, using the letters on the diagram in your answer.

r What are the advantages of this circuit**?**

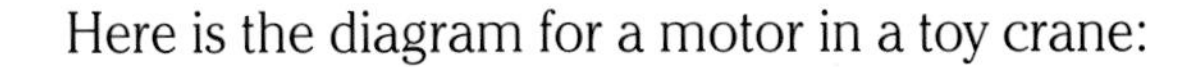

Here is the diagram for a motor in a toy crane:

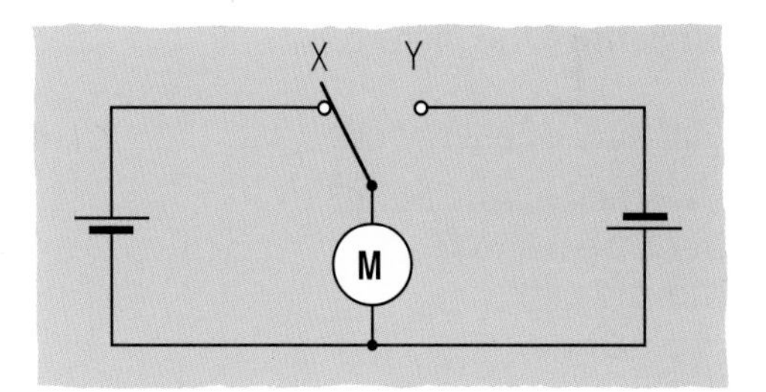

s What do you notice about the cells**?**

t What happens if the switch is moved from X to Y**?**

Here is a more complicated circuit:

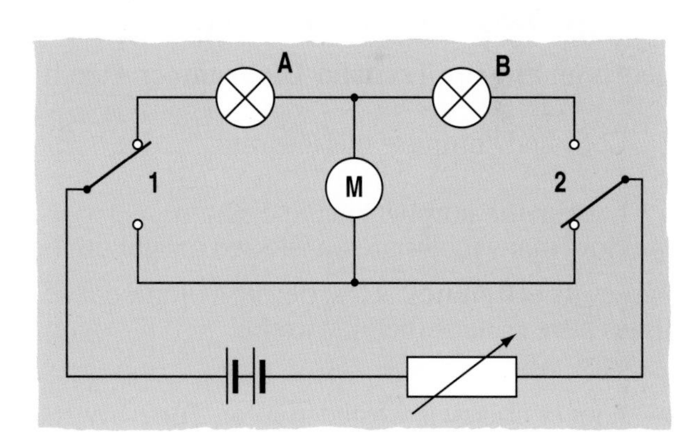

u Copy and complete this table.

Switch 1	*Switch 2*	*Bulb lit?*	*Motor turns?*
up	down	A	forwards
up	up		
down	up		
down	down		

v What would be the effects of varying the resistor**?**

Things to do

1 Draw a circuit containing a cell, a switch and 3 bulbs labelled A, B and C. Bulbs A and B are in series, and C is in parallel with A. The switch controls only bulb C. When all the bulbs are lit, which one is the brightest?

2 Draw a circuit with a 2-way switch, a cell, a bulb, and a motor so that in one position the bulb is on and in the other position the motor is on.

3 In these circuits, all the bulbs are the same:

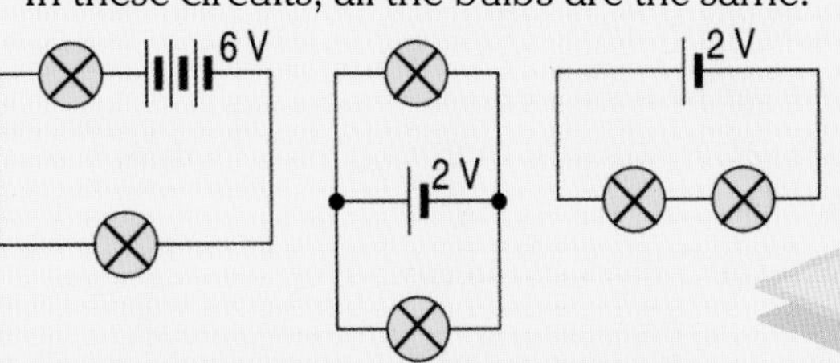

a) Which circuit has the dimmest bulbs?
b) Explain why this is, using the words:
current voltage resistance

Generating electricity

Learn about:
- efficiency
- generating electricity

▶ Julie is playing with her group:

In the picture there are 8 forms of **energy** labelled.

a Write down the names of these 8 forms of energy.

Nuclear energy is not shown here. Add it to your list.

b What is nuclear energy**?**

Some energy is called **potential** energy.

c Write down 2 examples of potential energy from the picture.

▶ What kind of energy has

d a stretched elastic band**?**
e a book on a shelf**?**
f a moving car**?**

Energy diagrams and efficiency

Here is an **Energy Transfer Diagram** for a torch:

.... energy stored in the .... **100 J**

.... energy in the wires

.... energy lighting up the room

.... energy heating up the torch and room **95 J**

g Copy and complete this diagram.

h Look at the numbers in the diagram. How many joules are transferred to light up the room**?**

We say its **efficiency** is 5%, because only 5 out of 100 joules have done something useful.

▶ A lot of energy is wasted in a car. For every 100 J of chemical energy in the petrol, only 25 J are transferred to kinetic energy. The rest just heats up the engine and the air.

i Draw an Energy Transfer Diagram for this, to scale.

j What is the efficiency of the car**?**

Although there is the same amount of energy afterwards, not all of it is useful.

This is summed up in the 2 laws of energy:

Law 1 The total amount of energy in the universe stays the same. It is 'conserved'. Energy cannot be created or destroyed.

Law 2 In energy transfers, the energy spreads out, to more and more places. As it spreads, it becomes less useful to us.

Making electricity

One way to make electricity is to use the energy of falling water.
Your teacher can show you this:

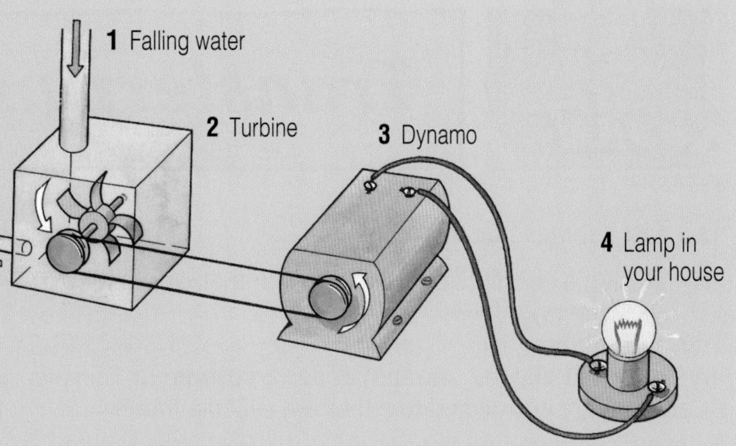

1 Falling water
2 Turbine
3 Dynamo
4 Lamp in your house

The falling water turns a turbine, and this turns a dynamo.

This is like a **hydro-electric power station**, where the water is stored behind a dam.

k What form of energy does the water in the dam have**?** (Hint: it is high in the mountains.)

l Copy and complete the energy flow-chart for this:

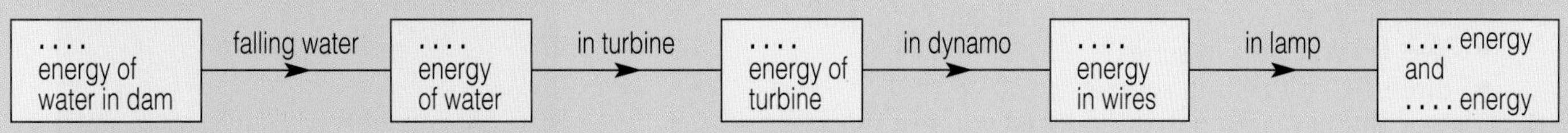

m How could you make electricity from energy in the ***wind*?**

▶ Another way to make electricity is to burn a fuel, which boils water to make steam. The steam then turns a turbine, and this turns a dynamo (a generator):

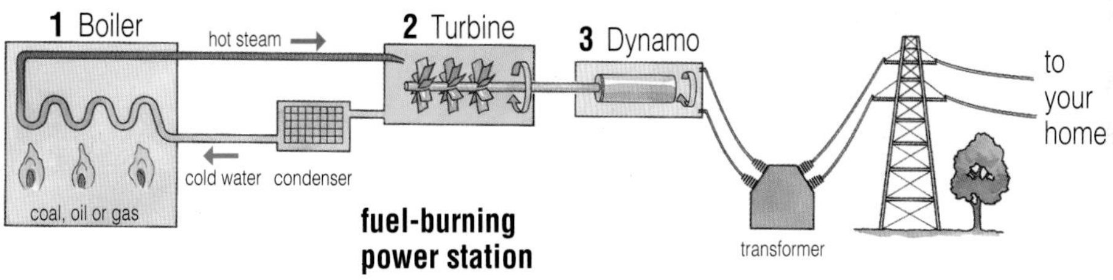

1 Boiler
hot steam →
2 Turbine
3 Dynamo
to your home
← cold water
condenser
coal, oil or gas
transformer
fuel-burning power station

n Draw a flow-chart for this power station, like the one in question **l** above.

o How could this be adapted to use nuclear energy**?**

▶ Fuel-burning power stations cause pollution.
They give out sulphur dioxide and carbon dioxide gases.

The sulphur dioxide dissolves in the rain to cause **acid rain**:

Trees killed by acid rain

p Carbon dioxide 'traps' the Sun's energy and keeps the Earth warm.
Why is this called the **greenhouse effect?**

q In your group, choose one way of generating electricity, and find out its effects on our environment.
Prepare to report your findings to the other groups.

Things to do

1 Copy and complete:
a) Energy is measured in ....
b) The 9 forms of energy are:
c) Lifting a stone gives it .... energy.
d) In any energy change, the total amount of .... before the change is always .... to the total amount of .... afterwards.
e) After the change, not all of the .... is useful.
f) Sulphur dioxide from power stations can cause .... rain, and carbon dioxide may warm up the Earth too much, because of the .... effect.

2 What are the energy transformations in
a) a Bunsen burner? b) a television?
c) a yo-yo? d) a hair-dryer?
e) a clockwork toy? f) an apple tree?

3 In a nuclear power station, nuclear energy heats water to make steam. Draw a flow-chart for it, like the one shown above.

4 In one second a lamp transfers 100 J, but only 4 J is light energy. 
a) Draw an Energy Transfer Diagram for it.
b) What is the efficiency of the lamp?

Transferring energy

Learn about:
- transferring energy
- power ratings
- the cost of electricity

a List 4 things that use a battery.

b Design a circuit diagram for a torch that would let you vary the brightness.

A single **cell** like the one shown has a **voltage** of 1.5 volts.
Cells that are connected together are called a battery.

c What voltage do you think you would get if you put
- two 1.5 volt cells in series?
- three 1.5 volt cells in series?

d Draw a circuit diagram that would let you test this.

A cell pushes electrons round the circuit so there is a current.
A bigger voltage gives a bigger push to the electrons.

In a torch, the current ***transfers energy*** from the cell to the bulb.
The higher the voltage, the more energy is transferred to the bulb.

The voltage across a component is also called the **potential difference** (or p.d.).

Chemical energy in a re-chargeable battery is transferred to an electric motor, where it becomes kinetic energy

Investigating voltage

Connect up this circuit:

Take care to connect the voltmeter correctly.

e Which components are in series?

f Which components are in parallel?

g Move the slider to vary the resistance in the circuit.
Look at the voltmeter and at the bulb.
What pattern can you see?

h Leave the slider in a fixed position.
- First measure the voltage across the bulb.
- Then move the voltmeter so it is connected across the resistor.
Measure the voltage across the variable resistor.
- Then move the voltmeter again, and measure the voltage across the battery.

What do you notice about your results?

i Repeat **h** with the slider in different positions.
What pattern do you find?
Can you explain it?

j Draw a circuit diagram of the circuit shown.

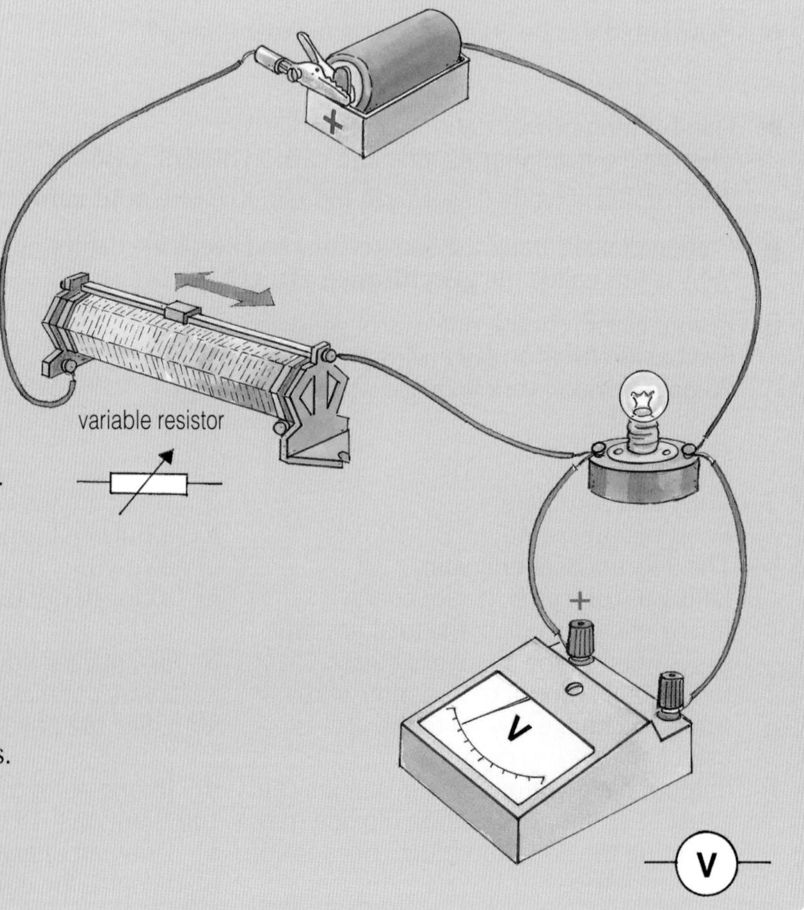

Energy in your home

Most of the electrical energy that you use in your home comes from the 'mains'. Mains voltage in the UK is 230 V.

k Why does this mean it is dangerous?

l Why does a mains plug have a fuse inside it? How does it work?

Different appliances in your home use different amounts of energy in each second. For example:

- the TV shown here transfers 100 joules in each second,
- the electric kettle transfers 2000 joules in each second.

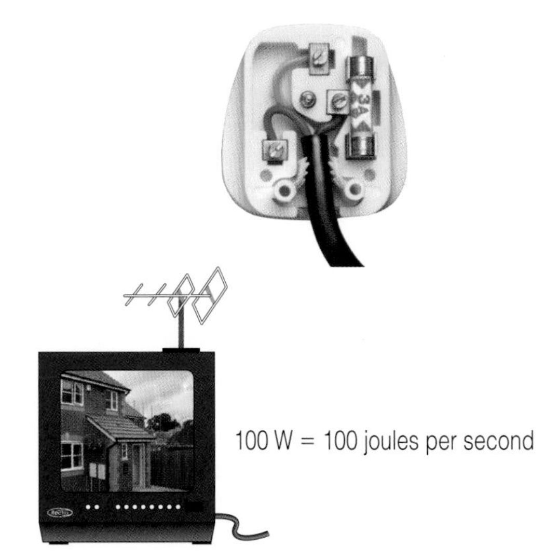

The **rating** of each appliance is usually marked on a label (often on the back or bottom of the appliance).

For example, the electric kettle is marked 2 kW (2 kilowatts).
2 kW = 2 kilowatts = 2000 watts = 2000 joules ***in each second***.
This is the amount of energy transferred from the power station to the water in the kettle.

m How many joules would the kettle transfer in 10 seconds?

2000 W = 2000 joules per second

Inside the electric kettle there is a wire with a lot of **resistance**. As the electrons are forced through this resistance, they heat it up. The electrical energy is changed ('transformed') to heat energy.

Comparing the cost

You (or your parents!) have to pay for the energy.
To run a one-bar electric fire for 1 hour costs about 8p.
It doesn't sound much, but it adds up.

And the real cost is more than 8p. It is also the effect on our environment (acid rain, global warming, loss of fossil fuels).

Your teacher will give you a Help Sheet. Use it to put some appliances in order, starting with the least expensive to run.

Design a leaflet for parents entitled 'How to get your children to save energy by turning the right things off'.

An electricity meter

Things to do

1 Copy and complete:

a) A battery pushes . . . . round a circuit. The size of the push is measured in . . . . by using a . . . .

b) Two 1.5 volt cells in series can give a voltage of . . . . volts.

c) In a torch, the current transfers . . . . from the . . . . to the . . . .
The higher the voltage, the more . . . . is . . . . to the bulb.

d) In an electric kettle, the electrical . . . . is changed to . . . . energy because of the . . . . of the wire.

2 Draw circuit diagrams of these:

a) A cell and a switch connected to 3 bulbs in series, with a voltmeter connected across one of the bulbs.

b) A battery of 2 cells in series with a variable resistor and 2 bulbs and an ammeter. One of the bulbs has a voltmeter across it.

c) A cell in parallel with 2 bulbs.
One of the bulbs is controlled by a switch. The other bulb is controlled by a variable resistor and this bulb has a voltmeter across it.

Physics at Work

Learn about:
- transferring and using energy

The diagram shows part of a **'mains' circuit** for a house:

a Are the lamps in series or in parallel?

b What is a fuse for?

c How does it work?

d What happens if fuse A 'blows'?

e How does the staircase circuit work?

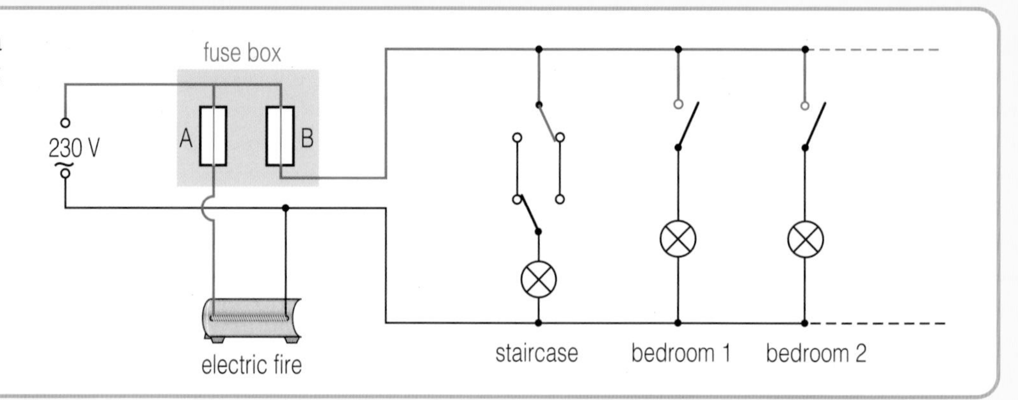

An **elephant** sometimes gets too hot.

f It has big ears. Explain how this helps it to cool down.

g Explain carefully why it sometimes uses its trunk to spray water over its back.
How many reasons can you think of?

This **marathon runner** has just finished a race.

h Why is she wearing a shiny plastic cover?

The **brakes** on this racing car are glowing red-hot.

i Why is this?

j What happens to this heat energy?

The diagram shows a **rocket-balloon**:

k When the balloon is blown up, what kind of energy does it have?

l Where did this energy come from?

m When the air is released,
- what happens to the balloon?
- what happens to the energy?

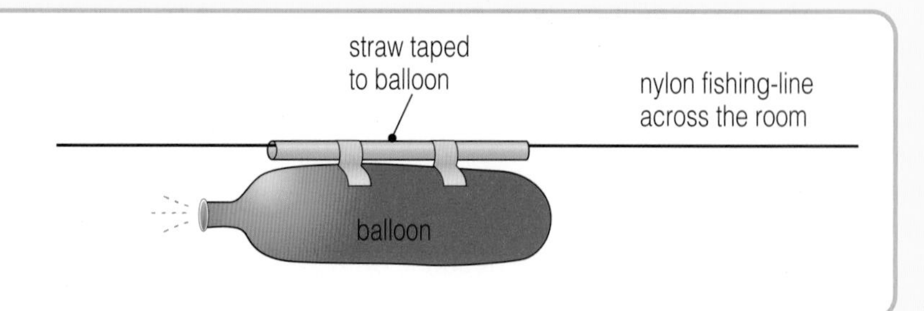

An ordinary filament **light bulb** is only about 5% efficient. So for every 100 joules of electrical energy, only 5 joules are transformed to light energy.

n What happens to the other 95 J**?**

An energy-saving bulb is also shown:
For every 100 J of electrical energy you get 25 J of light energy.

o What is its efficiency**?**

p Draw an Energy Transfer Diagram for each kind of bulb, to scale.

The diagram shows a side view of a **solar panel**.
It is using solar energy to heat water for a house.

q Which way is water flowing in the pipe**?**

r Why is the hot water outlet at the ***top*** of the tank**?**

s Explain how energy is transferred through the wall of the pipe to the water inside the tank. Use the idea of particles (atoms or molecules) in your explanation.

t Why is there a black surface in front of the pipe and a shiny surface behind it**?**

u Explain, step by step, how energy is transferred from the Sun to the hot water tap.

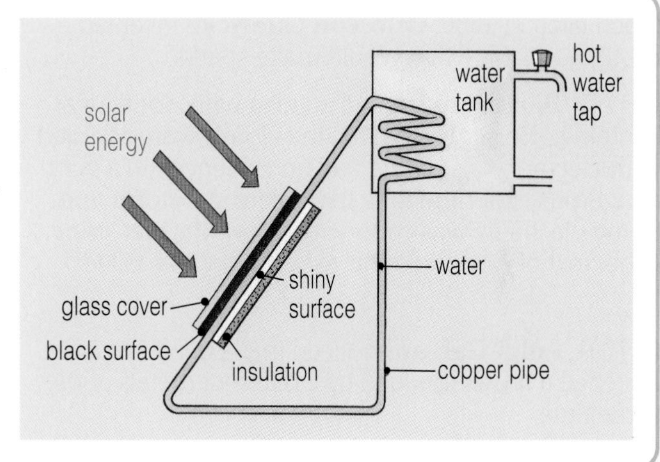

v Why are chicks covered with fluffy feathers**?**

w What happens if they get covered in oil**?**

x Draw an Energy Transfer Diagram for this **electric car**.

y It has re-chargeable batteries in it. Explain how its energy probably comes from the Sun.

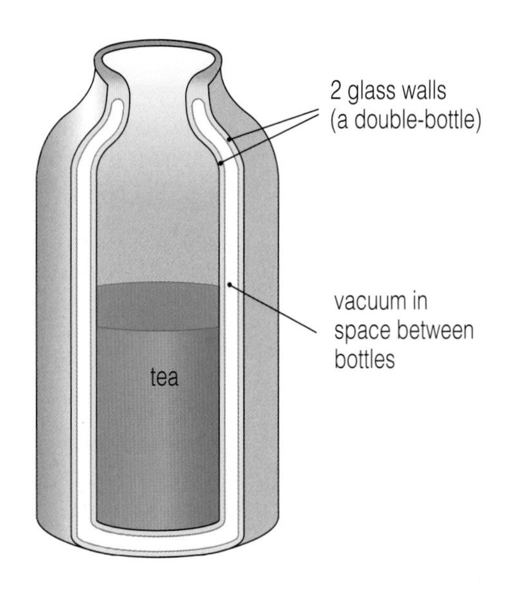

The cut-away diagram shows the *inside* of a 'thermos' **vacuum flask**, for keeping tea warm.

z How does this design stop heat transfer by conduction**?**

A How does this design stop heat transfer by convection**?**

Electricity & Magnetism

Learn about:
- early ideas
- how ideas change

Ideas about Electricity

2500 years ago, a Greek called **Thales** experimented with a piece of amber resin. He found that if he rubbed it, it would attract tiny pieces of paper, just as your comb can when you rub it on dry cloth. (Amber was called *elektron* in Greek.)

Much later, in 1660, **Otto von Guericke** invented a rubbing machine that would make sparks.

In 1752, **Ben Franklin** had an idea while looking at lightning. He suggested that the clouds were charged with electricity . . . but he had no evidence. In a very dangerous experiment he flew a kite in a storm and found electricity was conducted down the wet string. (The next person to do this experiment was killed!)

In 1791, **Luigi Galvani** noticed that a dead frog's leg twitched if it was touched by 2 different metals at the same time.
Alessandro Volta followed up this idea, and in 1800 he invented the first battery.
He used 2 metals (silver and zinc) with blotting paper (soaked in salt water) between the metal discs. He built a tower (a 'voltaic pile') of these discs, as you can see in the picture:
All the cells pushed the same way to give a bigger voltage, and the battery could deliver a current round a circuit. Scientists were astonished at this new idea.

Now that a steady current could be used, electricity was studied by many people, including **André Ampère** and **Georg Ohm**.

Volta shows his battery to Napoléon Bonaparte

Ideas about Magnetism

The first magnets were pieces of rock (iron oxide) that happened to be magnetised. They were called lodestones. The lodestone had been magnetised by the Earth's magnetic field.
About 1000 years ago, sailors began to use a lodestone as a compass.

William Gilbert was Queen Elizabeth I's doctor and he also investigated magnetism. In 1600 he published the results of his experiments.
He wanted to explain why a compass points North-South, so he built a model of the Earth with a lodestone inside, and used this to show that the Earth acts as if it has a bar magnet inside it.

Ideas about Electromagnetism

Some scientists suspected a link between electricity and magnetism, but could not find a way to show it.

Hans Christian Ørsted was convinced that there was a link, and tried many experiments, but it was hard to get a strong current and a strong magnet.
It wasn't until 1820 that he discovered, almost by accident, that a current passing through a wire can affect a compass needle that is placed nearby.
This was the first ammeter, because the bigger the current, the more the compass needle is moved.

In 1821, **André Ampère** used a coil to make the first electromagnet.

Hans Ørsted discovers electromagnetism
Can you see his battery and his compass?

Michael Faraday came from a poor home, but got a job as a lab assistant. He soon became the greatest experimenter in Physics.
In 1821 he invented the electric motor.

He also invented the idea of using 'lines of force' (or 'lines of flux' or 'field lines') to give a picture of a magnetic field. This is a very important idea. It is used by modern scientists to picture other fields, such as gravitational fields.

By 1831, after many experiments, Faraday discovered the transformer and the generator. This is a machine for making electricity (it is sometimes called a dynamo or an alternator). You may have one on your bicycle. It converts kinetic energy to electrical energy. It is vital to modern society.

Michael Faraday

In 1879 **Thomas Edison** in America and **Joseph Swan** in England developed the electric light bulb, independently. Soon, power stations were being built to deliver electricity to factories and homes, and provide street lighting.

In 1897 **J.J.Thomson** did an important experiment with a cathode ray tube (a kind of TV tube). By using a magnetic field to bend the beam of current inside the tube, he showed that a current is really a flow of negative electrons.

Thomas Edison's first successful carbon filament lamp

Questions

1 Where do you think the word electricity came from?

2 Draw an Energy Transfer Diagram for
a) a motor and b) a generator.
In what ways are they opposites?

3 a) Draw a labelled diagram of one of Volta's cells.
b) One of his cells gives a voltage of 1.8 V.
How did he get a voltage of 9 V?

4 The nerves in your body use electricity. How does Galvani's experiment support this idea?

5 Use the dates shown above to draw a time-line.

6 How did Edison's light bulb change people's lives?

7 The article says the generator 'is vital to modern society'. Explain what this means, with examples.

8 Choose one of the scientists and find out more details of his life.

9 Scientists work in many different ways. Use the article above to list as many ways as you can.

Questions

1 Look at this photo of a down-hill skier:

a) What kind of energy does he have at the top of the hill?
b) What kind of energy is it changed into, as he moves down-hill?
c) There is some friction between his skis and the snow. What can you say about the temperature of the skis as he slides down?
d) What other friction force does he feel? Why is he crouching?
e) What has happened to all the energy when the race is finished?

2 This circuit is used to dim a light:

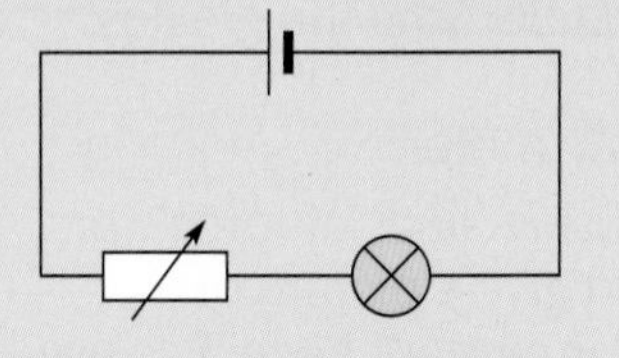

While the light is being dimmed, what happens to
a) The current in the lamp?
b) The resistance of the variable resistor?
c) The voltage across the lamp?

3 The diagram shows an electrical circuit:

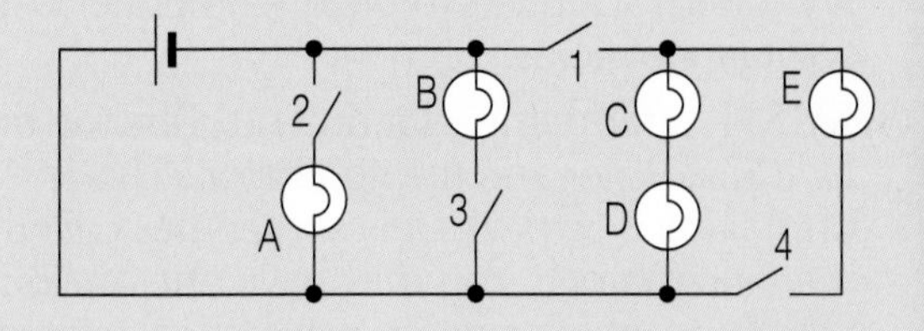

a) What happens if only switch 1 is closed?
b) What happens if only switches 3 and 4 are closed?
c) How would you light bulbs A, C and D?
d) If all the switches are closed, which bulb(s) are dimmest? Why?

4 In the circuits shown, all the cells are identical, and all the bulbs are identical:

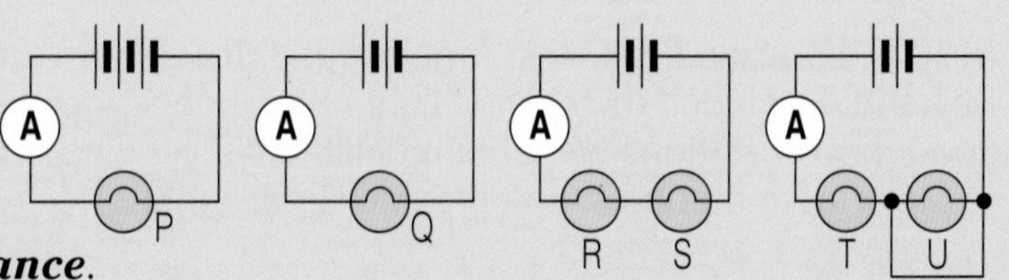

a) Which bulb would be brightest? Why?
b) Which bulb would not be lit? Why?
c) Which ammeter would show the lowest reading? Explain why, using the words ***voltage*** and ***resistance***.

5 Danny connected a battery in series with a switch, an ammeter, a variable resistor and a coil of wire. Then he connected a voltmeter to measure the voltage across the coil.

a) Draw a circuit diagram of his circuit.

The table shows his results as he altered the variable resistor:

Current (A)	0.4	0.7	1.2	1.5	1.9	2.4	2.8
Voltage (V)	1.9	2.9	4.9	6.3	8.0	10.1	11.8

b) Plot a large graph of the current against the voltage, drawing the line of 'best fit'.
c) Which result(s) do you think could be wrong and should be checked?
d) What would be the current if the voltage was 8.4 V?
e) What conclusion(s) can you draw from the graph?

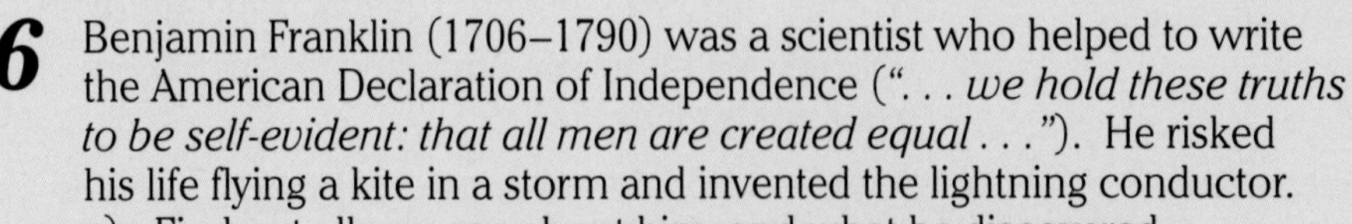

6 Benjamin Franklin (1706–1790) was a scientist who helped to write the American Declaration of Independence ("*. . . we hold these truths to be self-evident: that all men are created equal . . .*"). He risked his life flying a kite in a storm and invented the lightning conductor.
a) Find out all you can about him, and what he discovered.
b) How have our ideas about electricity changed since he was alive?

Ben Franklin

Gravity and space

Everything you do is affected by the pull of gravity, and we need this force to live on Earth.

On a larger scale, gravity is the force that keeps the planets in our Solar System, and keeps the beautiful rings round Saturn:

The pull of gravity

Learn about:
- mass and weight
- gravitational force

The diagram shows a boy standing in 4 different countries on Earth, and dropping a ball:

a Copy the diagram and draw in the path of the ball in each case.

b What is the name of the force that pulls the ball?

c Which way is "up"?

A mass of 1 kilogram on Earth weighs 10 newtons:
We say that the gravitational field strength is 10 newtons per kilogram (10 N/kg).

d What is the weight of a man whose mass is 60 kg?

e The mass of this book is 0.6 kg. What is its weight?

newtons
0
10
1 kg

On Earth,
1 kg weighs 10 N

On the Moon a kilogram would weigh *less*.

f Why do you think this is?

The pull of gravity on the Moon is about $\frac{1}{6}$th of the force of gravity on Earth.

g How much would the man in **d** weigh on the Moon?

h How much would this book weigh on the Moon?

i What is the mass of this book on the Moon?

The pull of gravity is different on different planets, as shown in the table:

j Decide what it means, then copy and complete it.

k Can you see a pattern in the numbers?

Planet:	Earth	Mars	Jupiter	Pluto
Gravitational field strength in N/kg	10	4	26	0.5
Weight of 1 kg bag of sugar				

Lift-off

To leave the Earth, a rocket motor has to overcome the pull of gravity.
A Saturn V Moon rocket has a mass of 3 million kg.
The thrust of its motor at lift-off was 32 million N.

l What was the weight of the rocket on Earth?

m Draw a diagram of the 2 main forces on the rocket.

n What was the resultant force at lift-off?

As the rocket leaves Earth, the pull of gravity decreases (unless it goes near the Moon, the Sun, or a planet).

o What would be the weight of the rocket if it went deep into space, far away from any planets or stars?

A Saturn V Moon rocket

An astronaut experiments on the Moon

The Solar System

The Solar System consists of the Sun and the 9 planets:

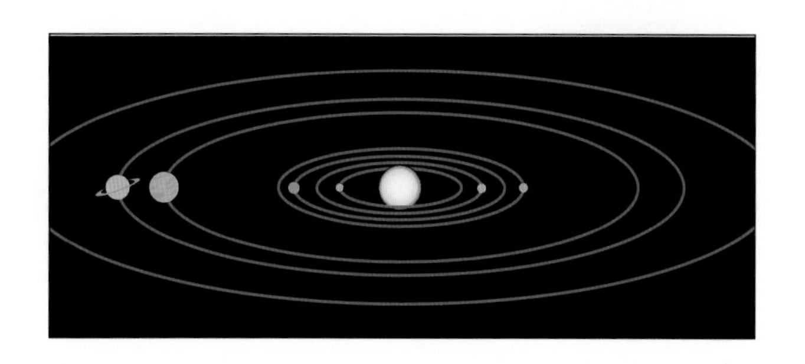

p Which object in the Solar System has the most mass**?**

q Which object in the Solar System exerts the biggest gravitational pull**?**

r What are the names of the 9 planets, in order**?**

The planets move in orbits which are almost circles.

Here is a diagram of a cork on a piece of string, being whirled round in a circle:

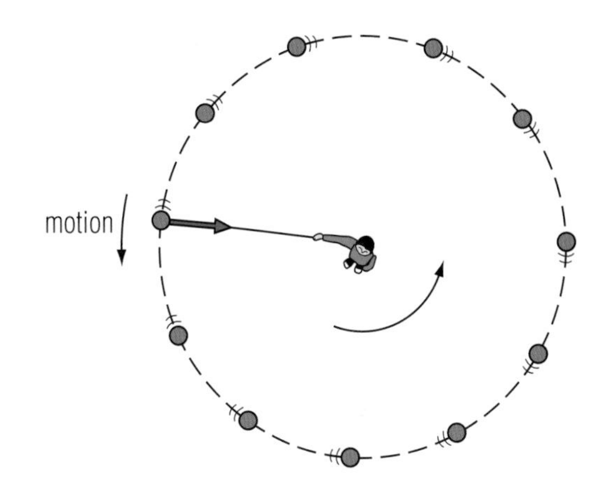

It is rather like a planet in orbit.

s What is the force that keeps the cork moving in a circle**?**

t What is the force that keeps a planet in its orbit round the Sun**?**

u What would happen to the cork if the string suddenly broke**?**
Discuss this in your group, and draw a diagram to show what you think happens to the cork.

The Moon moves in an orbit round the Earth. It is a natural **satellite**.

v Explain why the Moon moves in an orbit round the Earth.

w Why does the Moon look different to us at different times of the month**?**

x Why does the Moon shine brightly even though it is very cold**?**

y Why do you think the Moon has more craters than Earth**?**

Things to do

1 Copy and complete:

a) The . . . . of an object stays the same everywhere, but the . . . . of the object always depends on the pull of . . . .

b) Compared to Earth, the pull of . . . . is *more/less* on the Moon, and *more/less* on Jupiter.

c) The planets are held in their . . . . by the gravitational pull of the . . . .

2 An astronaut travels from Earth to the Moon. Explain how (i) her mass and (ii) her weight change during the journey.

3 Use an encyclopedia or the internet to research and write about the main events in the human exploration of space. Some of the people include Yuri Gagarin, Valentina Tereshkova, and Neil Armstrong.

4 The diagram shows a firework rocket:
As it flies through the air, there are 3 forces on it. Copy the diagram.

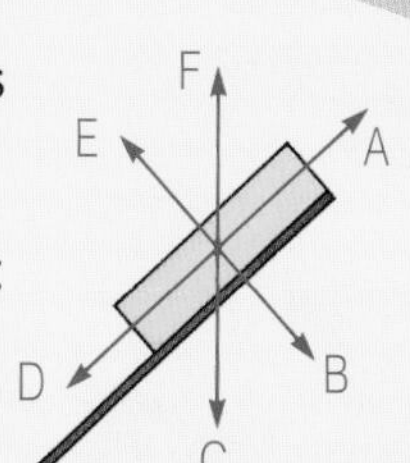

a) Which 3 arrows show the 3 forces?

b) Label the 3 arrows, using these words:
weight ***air resistance*** ***thrust***

c) What can you say about these forces when the rocket is just taking off?

d) Why does the rocket come back down?

9J2 Ideas about gravity

Learn about:
- how ideas develop
- gravity and space

350 B.C. **Aristotle** thought that heavy objects contained a substance called gravity.

He thought that light objects had a substance called levity.

For 2000 years, people believed this!

Then, in 1589, **Galileo** shocked people by actually doing experiments!

He is supposed to have used the leaning tower at Pisa:

They both have exactly the same acceleration.

In 1666, **Isaac Newton** was thinking about gravity:

Isaac calculated how much the Earth pulled on the Moon.
He found that the farther the distance, the weaker the force of gravity.
At twice the distance, the force is $\frac{1}{4}$.
At 3 times the distance, the force is $\frac{1}{9}$.

Newton took his ideas further . . .

Sir Isaac Newton
1642–1727

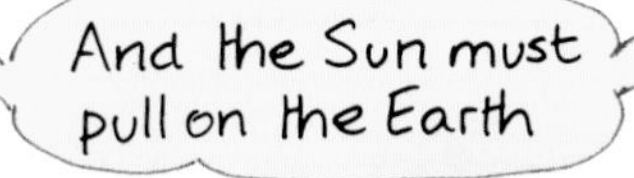

So the Solar System is held together by the force of GRAVITY!

Isaac worked out a formula:

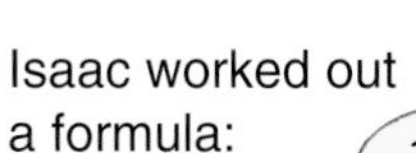

The force between 2 bottles on a shelf is only $\frac{1}{200\,000\,000}$ of a newton

– but the pull of the Earth on you is about 500 N
– your **weight**.

He published his ideas in a famous book.

Newton's ideas seemed to work perfectly, until . . .

1916 **Albert Einstein** thought about the effect of gravity on space itself.

Let's think of space as like a rubber sheet.

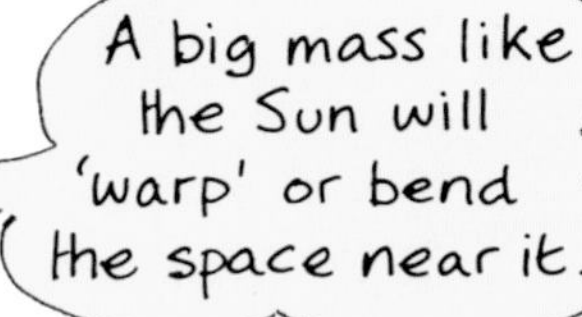

Like this marble a space-ship will travel in a curve. BUT — so also will a ray of light.

So, light is affected by gravity.
(An experiment proved it later.)

If the gravity of a star is very, very strong, it will pull so hard that the light from the star cannot escape . . .
. . . it is a **black hole**!

Things to do

1 Use the idea of gravity to explain:
a) Why is an astronaut lighter on the Moon than on the Earth?
b) What evidence is there that Saturn exerts a gravitational pull?
c) Which object has the greatest pull in our Solar System?
d) Why do the planets not travel in straight lines?

2 The diagram shows 4 cups on the Earth's surface. Copy the diagram and draw each cup half-filled with water.
Which way is 'up'?

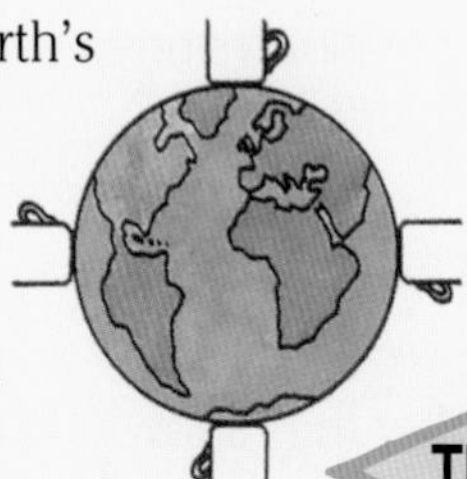

3 An astronaut on the Moon dropped a feather and a hammer at the same time. They hit the ground together. Why does this happen on the Moon but not on Earth?

Rockets and satellites

Learn about:
- launching satellites
- uses of satellites

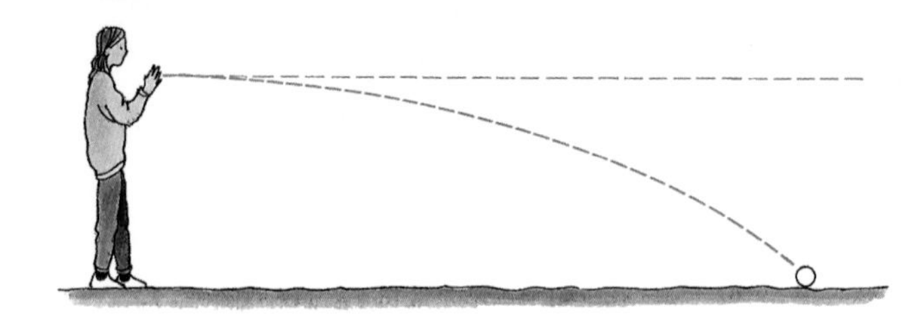

Anna throws a ball horizontally:
The ball moves in a curve.

a Why does the ball not move in a straight line?
b What happens if she throws it harder?
c What happens if she fires a bullet horizontally?

Escaping from the Earth

Imagine firing a gun from the top of a very high mountain:

The bullet will fall back to Earth, just like a ball. It is pulled by gravity. This is shown at **A** in the diagram:

What if the bullet could be fired faster? It would go farther, but still fall to Earth. Look at **B** and **C**.

Suppose the bullet could be fired even faster, at 25 times the speed of sound. It would still fall towards Earth but, because the Earth is curved, it would stay the same height above the ground. Look at **D**.

It is now in orbit. It is a **satellite**.

d What can happen if it travels even faster than **D**? Draw a diagram of this.

A
B
C
D it is still falling, but does not hit the ground

You can't do this with a gun, but you can with a **rocket**.
It needs a speed of about 29 000 km per hour (18 000 mph).

A satellite moves very fast, but it can seem to be standing still !
If the satellite is put at just the right height and speed, it takes ***24 hours*** to go round the Earth once.
This is the same time as the Earth takes to spin round once – so the satellite appears to stay over one place !
This is called a **geo-stationary** satellite.

Rockets

- Blow up a balloon and then let go.
  What happens? Why?

  As the air rushes backwards, the balloon moves forwards. A rocket works in the same way.

- Your teacher may show you a water rocket, using compressed air.
  What happens when the water is forced out? Why?

The space-shuttle taking off

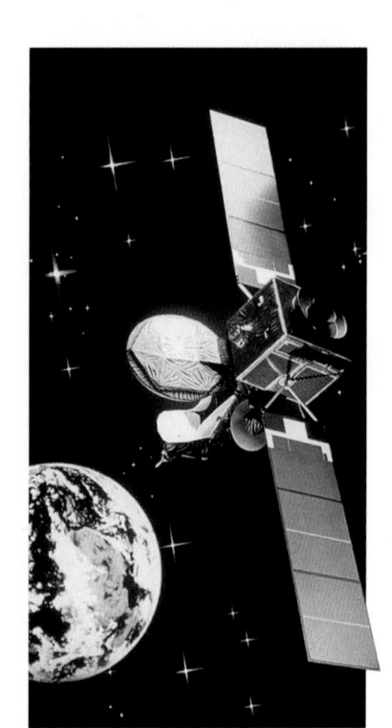

A communications satellite

Using satellites

Communications satellites can be used to send telephone messages or TV pictures:

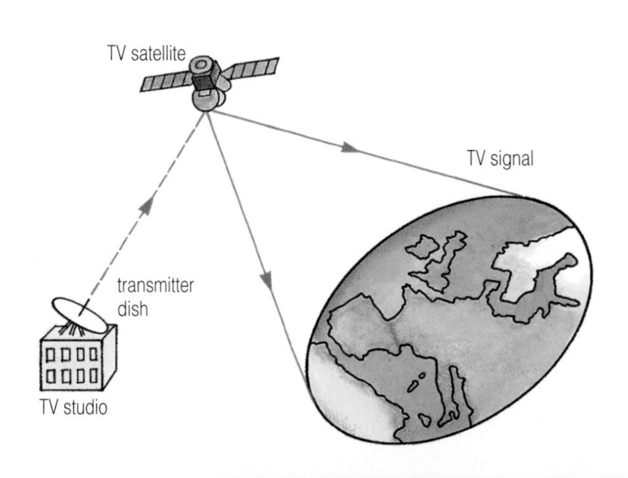

We use a geo-stationary satellite. It moves with the Earth, and stays over the same place all the time.

You can use a 'dish' aerial on your house to get the signal.

e Why are the dishes used in the north of Scotland bigger than those used in the south of England?

Weather satellites are useful. They take photos of the weather, and send the pictures back to Earth by radio.

f Look at the weather picture shown below. Write down all the conclusions you can draw from it.

Land survey satellites are very useful. These are lower satellites that take photos in great detail. They can warn of forest fires and water pollution; show healthy and diseased crops; and help us to find oil.

Military satellites are used for spying.

Navigation (GPS) satellites are used by ships, planes, taxis, walkers. You can find where you are, anywhere on Earth, to within 10 metres !

Astronomy satellites carry telescopes. They can take very sharp pictures of planets, because they are outside the Earth's atmosphere.

The Earth seen from a satellite

A weather picture of the UK/Europe. What conclusions can you draw?

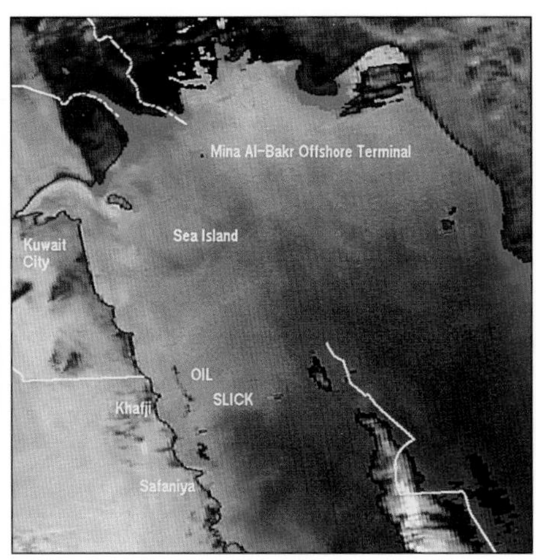

An oil slick polluting the sea. Which part of the world is this?

This is London. Can you see the bridges over the river Thames?

Things to do

1 Copy and complete:

a) All objects are pulled to the Earth by . . . .

b) If it travels fast enough, a satellite can stay in . . . . even though it is falling towards Earth all the time.

2 Why do satellites not need to be streamlined?

3 Explain in your own words, and with a diagram, why a satellite stays in orbit.

4 'GPS' stands for Global Positioning System. The photo shows a portable GPS receiver:
Use an encyclopedia or the internet to find out about it. What use is it, and how does it work?

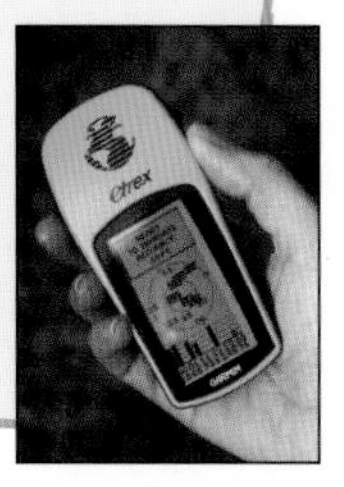

Changing ideas

Learn about:
- early ideas about forces and the Solar System
- how ideas change

Ideas about forces

One of the earliest scientists was a Greek called **Archimedes**.
In about 260 BC he worked out the law of levers (also called the principle of moments).
He also discovered the idea of an upthrust in water, and why things float.
He discovered this in his bath, and was so excited that he ran naked into the street shouting "Eureka!" ("I have found it!").

Another famous Greek called **Aristotle** wrote about forces (see lesson 9J2). He believed, wrongly, that an object will only keep moving as long as we exert a force on it.
In everyday life on Earth this idea often appears to be true, because friction forces try to slow down a moving object.

Aristotle also said that heavy objects always fall faster than light objects. This often appears to be true, because air resistance slows down a falling leaf more than a rain-drop.

It wasn't until 2000 years later, in 1589, that **Galileo** actually did some experiments on moving objects, to test these ideas.
The traditional story says that he used the leaning tower in Pisa for his experiments, and dropped metal balls from the top of the tower.
In fact, in his experiments he rolled the balls down an inclined plane (a ramp).

Galileo Galilei 1564–1642

Galileo used heavy balls and light balls, and timed how long they took to travel down the ramp.
He showed that a force causes a ***change*** in an object's motion.

Isaac Newton was a great mathematician who had an enormous effect on ideas in Science.

He worked out 3 laws about forces, called the 3 laws of motion, which he published in 1687.
He also worked out and published his law of gravitation (see lesson 9J2).
These laws let us calculate the effects of forces.

Sir Isaac Newton 1642–1727

Forces are needed:
- to change an object's speed,
- to change an object's direction.

If there is no (resultant) force on an object, then the object just keeps on moving at the same speed, in a straight line.

Ideas about the Solar System

Once upon a time, people believed the Earth was flat.
But by 350 BC the Greek thinker **Aristotle** had good arguments for a round Earth. (For example, when a ship sails away, in any direction, you see the masts last of all.)

In 280 BC another Greek, **Aristarchus of Samos**, tried to measure the size of the Sun, and suggested that perhaps the Earth moved round the Sun.

However most people believed that the Earth was at the centre of the universe. They believed that the Sun was just a ball of light, and that the Earth was much heavier.
They believed that the Earth stayed still, and everything moved round it. This is called the 'geocentric' theory.

As time went by, astronomers observed the planets more and more carefully. They found it harder and harder to explain what they saw.

It wasn't until 1507 that a Polish monk called **Nicolaus Copernicus** took a fresh look at the problem.
He saw a simpler way to explain the observations.
He suggested that the ***Sun*** is at the centre and the planets move round it. This is called the 'heliocentric' theory.
This idea was very strongly opposed by the Church.

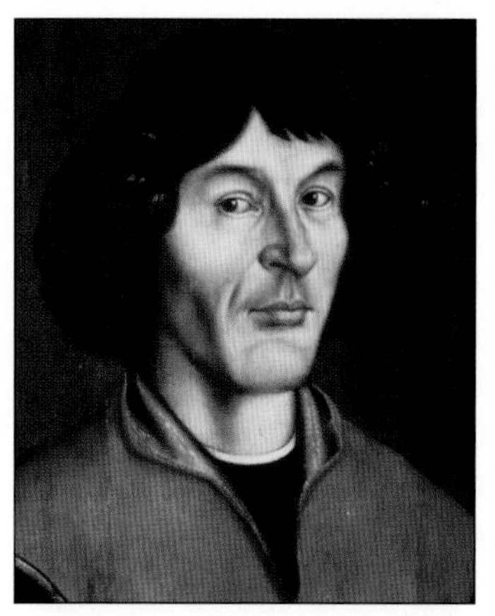

Nicolaus Copernicus 1473–1543

In 1609, **Galileo** used a telescope to observe Jupiter and its moons. What he saw convinced him that Copernicus was right.
This led to Galileo's famous argument with the Church, when he was threatened with torture and forced to retract his ideas.

Since 1570, **Tycho Brahe** had been making accurate observations of stars and planets.
Johannes Kepler looked at this data in detail. By 1609 he worked out that the planets must be moving in **ellipses**, not the "perfect circles" that everyone had believed.
He discovered 3 laws that described how the planets moved.

Later, **Isaac Newton** used the idea of gravity (see lesson 9J2) to explain ***why*** the planets move in elliptical orbits, to form the Solar System. He also explained why the Moon is held in orbit round the Earth.

Johannes Kepler 1571–1630

Questions

1 Sketch a rubber duck floating in Archimedes' bath and mark on it:
a) its weight, b) the upthrust of the water.

2 Think about Galileo's experiments with a ramp.
a) If you release a ball on a ramp, does it travel at constant speed? Describe its motion carefully.
b) Does a ball roll faster down a steep ramp or a shallow ramp? Can you explain why?
c) To reduce the effects of friction, would it be better to use a heavy ball or a light ball? Why?

3 Explain carefully how looking at a ship gives you evidence for a round Earth.

4 Give some reasons why you think people in the past "believed that the Earth was at the centre of the universe".
Then list some arguments against this model.

5 A space probe, far away in space, travels at a steady speed without using its engine. Why is this?

6 Choose one of the scientists and research his life.

Questions

1 The table shows the masses and weights of some objects on Earth, where 1 kg has a weight of 10 N.

a) Copy and complete the table.

b) On Mars, the weight of 1 kg is just 4 newtons. Add a row to your table to show the weights on Mars.

Object:	book	chair	girl	man	car
Mass in kg	0.6		30		1000
Weight on Earth in N		100		700	

2 Choose ***one*** of the following topics, and then use an encyclopedia or the internet to research it.
Write an explanation of what you find, using diagrams where possible.

a) Some satellites have a geostationary orbit, and some have a polar orbit. Why is this?

b) Find out the main events in the development of satellite technology. Draw a time-line to show them.

c) How was the Moon formed? There have been 2 main theories – what are they and which one has the best evidence?

3 Choose ***one*** of the following topics to write about.

a) Research, and then draw a time-line of the main events in the human exploration of space.

b) Research and write about the difficulties caused by low-gravity conditions on a space station.

c) Write a story about what would happen on Earth and throughout the Universe if gravity started to go weaker and weaker.

4 The diagram shows the journey (A, B, C, X, D) of an astronaut from the Earth to the Moon.
(The diagram is not to scale: the Moon should be farther away.)

At point X the gravitational pull of the Moon is equal to the gravitational pull of the Earth.

Write a story describing the journey, and explain at each stage how his mass and his weight are affected.

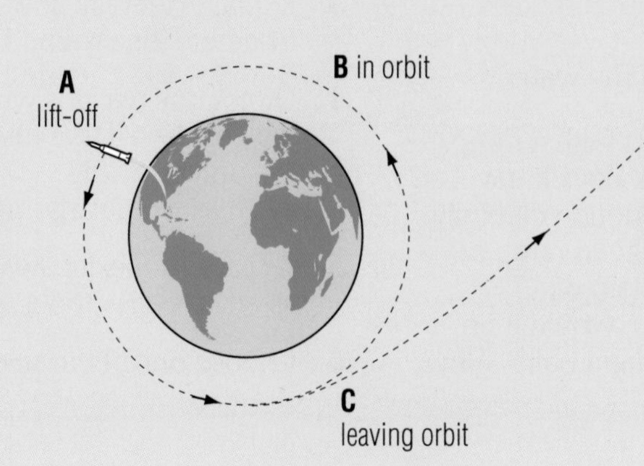

Speeding up

Everything that you do needs a force – a push or a pull.

You have already investigated forces, using a spring-balance to measure them (in newtons).
You found out about weight and friction. You investigated floating and sinking.

In this unit, we'll look at forces as they act on moving objects, to speed them up or slow them down.

Balanced forces

Learn about:
- resultant force
- balanced forces
- structures

Pushes and pulls are ***forces***.
The pictures show some forces, with their sizes measured in ***newtons* (N)**.

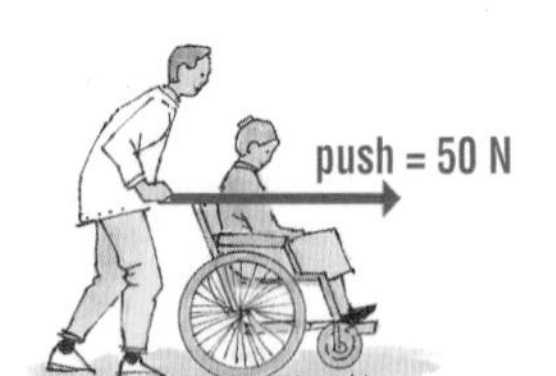

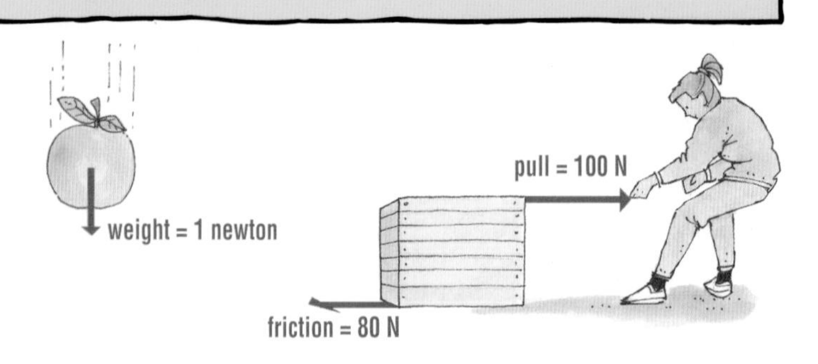

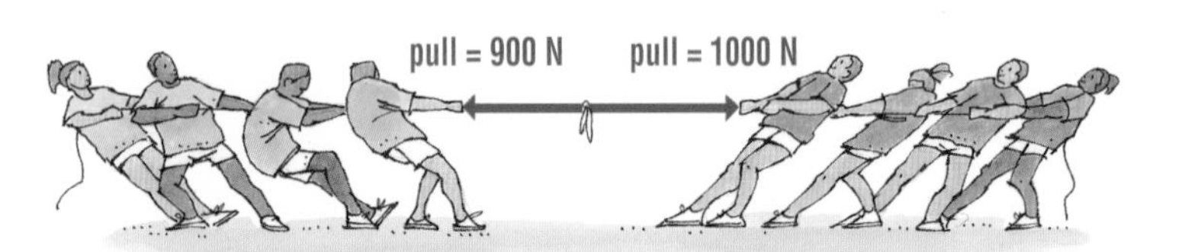

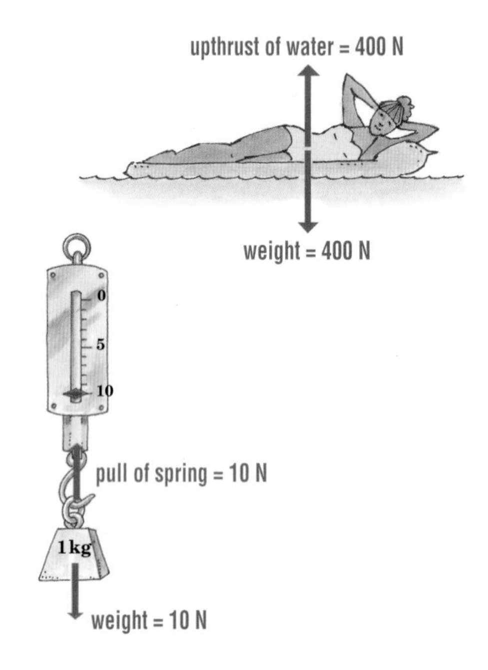

a Which is the biggest force shown in these diagrams?
b Which is the smallest?
c Which team is winning the tug-of-war? Which way is the rope moving? How do you know?
d How can you tell that the woman is moving the crate? How big is the ***resultant*** force on the crate?

In two of the diagrams the forces are ***equal*** and ***opposite***.
We say they are **balanced** forces.

e Which two diagrams show balanced forces?
f How do you know that the girl is floating and not sinking?
g What is the reading on the scale of the force-meter? What is the weight of a mass of 1 kilogram?

In each of the diagrams below, the forces are **balanced**.
Sketch or trace the drawings, and then for each one:
- label the size of the other force, in newtons,
- label the kind of force it is, choosing from:
  weight friction upthrust

h

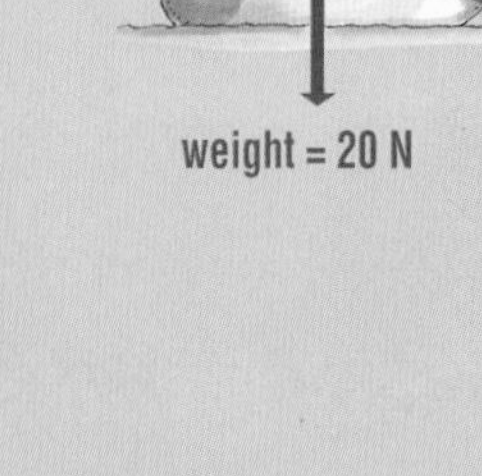

i

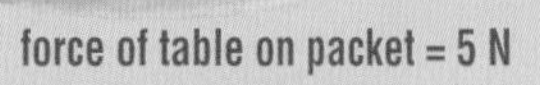

j

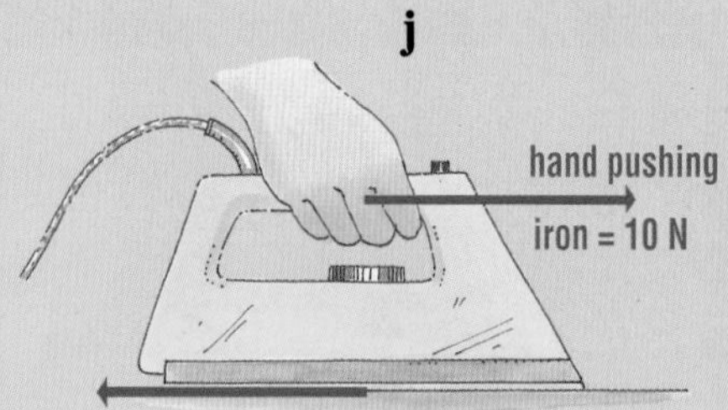

k

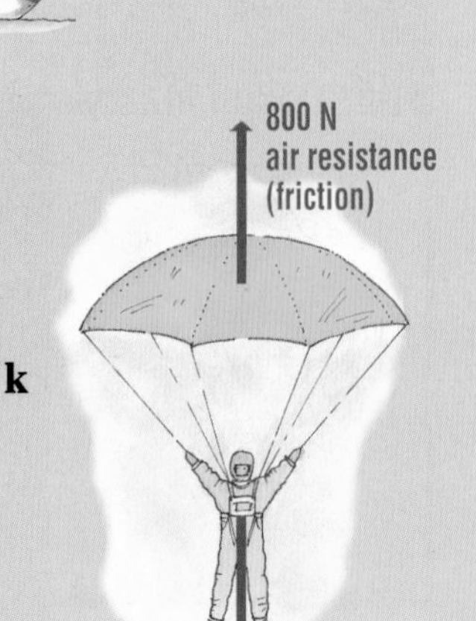

Sir Isaac Newton, 300 years ago, stated a scientific Law about balanced forces:
If the forces are balanced, the object either
- stays still (like the cornflakes packet)

or
- if it is moving, it continues to move at a steady speed in a straight line (like the parachute).

Structures

Here are some photos of **structures**:

Sometimes you can see the structure.
For example, in a bridge or a crane, a fence or a tree, or a bicycle.
Sometimes the structure is hidden.
For example, the beams in the roof of your house, or the skeleton in your body.

A structure can be designed by an engineer.
The structure must be strong enough to withstand the forces on it.
The forces in the structure must be ***balanced*** forces.

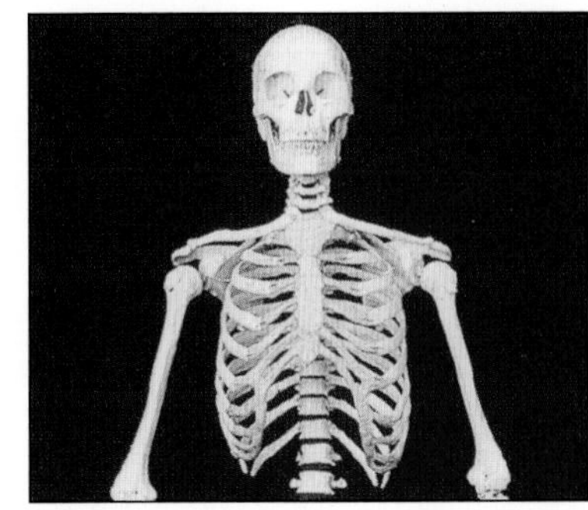

An engineering challenge!

Design and make a structure strong enough to support a 10 gram object as high as possible above the table.
You are given only:
- a 10 g object
- 20 straws
- 50 cm of sellotape. No more!

Your structure must be able to support the object for at least 30 seconds.
Try it!

- Who can build the tallest successful tower**?**
- Draw a labelled sketch of your tower.
- Look at the highest towers: how many triangles of straws can you count**?** Triangular shapes help to make a structure firm and rigid.

The straws you used are hollow tubes. A tube is stronger than a solid bar of the same weight. Why is a bike made from tubes**?**
Tubes are found in animals (e.g. a bird's bones) and in plants (e.g. the stem of a dandelion).

Things to do

1 Copy and complete:
a) Pushes and pulls are . . . .
b) When the forces on an object are equal and opposite, we say they are . . . .
c) Sir . . . . Newton's first Law is: if the forces on an object are . . . . , then
- if the object is still, it stays . . . .
- if the object is moving, it continues to . . . . at a steady . . . . in a . . . . line.

d) Structures are usually stronger if they are built of . . . . shapes.
e) A tube is . . . . than a solid bar of the same weight.

2 List all the structures that you can see in the classroom (or your home).

3 List all the structures that you can see on the way home.

4 Explain, with a diagram, why a bicycle frame is a strong structure.

5 Use the things you have learned to design a very thin tall tower for a TV transmitter. Draw a labelled diagram.

Moving at speed

Learn about:
- calculating speeds
- measuring speed

Some things can move fast. Other things move slowly. They have a different **speed**.

Suppose the horse in the picture has a constant speed of 10 metres per second.
This means that it travels 10 metres in every second.

The speed can be found by:

$$\textbf{Average speed} = \frac{\textbf{distance travelled} \text{ (in metres)}}{\textbf{time taken} \text{ (in seconds)}}$$

Travelling at 10 m/s

Speed can also be measured in miles per hour (m.p.h.) or kilometres per hour (km/h).

Example
In a race, this girl runs 100 metres in 20 seconds.
What is her average speed?

Answer

$$\text{Average speed} = \frac{\text{distance travelled}}{\text{time taken}}$$

$$= \frac{100 \text{ metres}}{20 \text{ seconds}}$$

$$= \underline{5 \text{ m/s}}$$

(This is about 10 m.p.h.)

This is her ***average*** speed because she may speed up or slow down during the race.
If she speeds up, she is ***accelerating***. If she slows down, she is ***decelerating***.

▶ Copy out this table, and then complete it.

	Distance travelled	*Time taken*	*Average speed*
a	20 m	2 s	
b	100 m	5 s	
c	2 m		1 m/s
d		10 s	50 m/s
e	2000 km	2 h	

▶ Now match the speeds in the table with these objects. Which is which?
Add the names to the first column of your table.

Investigating speed

Investigation 1

Investigate how fast you can • walk • run

You can use the formula (opposite page) or you may be able to measure the speed using a computer and a ***motion sensor***.

- How many measurements will you make?
- How will you record your measurements?
- Check your plan with your teacher before starting.
- Evaluate a method using data-logging compared with hand-timing using a stop-watch.

Investigation 2

Sam says, ***"I think people with long legs can always walk faster than people with short legs".***

- Plan a way to investigate this, and write a report explaining how you would do it.
- Check your plan with your teacher and, if you have time, do the investigation.
- Display your results on a graph and comment on any trends.

Investigation 3: A speed trap

Cars often drive fast past schools, and this is dangerous.

Plan an investigation ***to find out how fast the cars are travelling on a road near your school***.

Do not do this investigation unless your teacher agrees it is safe.

- What distance will you use?
- How will you start and stop your clock accurately?

If you do the investigation, find the speeds of 10 cars.

- Were all the cars travelling within the speed limit?
- Imagine you are a newspaper reporter. Use the results of your investigation to write a report for your newspaper.

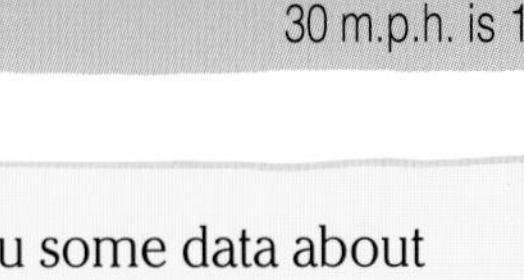

30 m.p.h. is 13 m/s

Things to do

1 Copy and complete:

a) The formula for speed is:

b) The units for speed are m/s (. . . . per . . . .) or km/h (. . . . per . . . .) or m.p.h. (. . . . per . . . .).

c) If a car speeds up it is . . . .
If it slows down it is . . . .

2 Imagine you are in a car travelling through a town and then along a motorway. Give an example of where your car might have:

a) a high speed but a low acceleration,

b) a low speed but a high acceleration.

3 A boy jogs 10 metres in 5 seconds.

a) What is his speed?

b) How far would he travel in 100 seconds?

4 The table gives you some data about 4 runners:

Name	*Distance (metres)*	*Time taken (seconds)*
Ali	60	10
Ben	25	5
Chris	40	4
Dee	100	20

a) Who ran farthest?

b) Who ran for the shortest time?

c) Who ran fastest?

d) Which people ran at the same speed?

5 A dog runs at 10 m/s for a distance of 200 m. How long did it take?

Can we go faster?

Learn about:
- air resistance (drag)
- streamlining
- resultant force

Air resistance

We are surrounded by air. When anything moves, the air rubs against it and slows it down. This sort of friction is called **air resistance** (or **drag**).

Air resistance is not usually a problem when we are just walking about, because we are not moving very quickly.
However it does become a problem if we want to go faster.

▶ Look how the shape of a motor car has changed over the last century.

a Which shape produces least air resistance?

▶ Modern cars and aircraft are designed to be as **streamlined** as possible. Scale models are tested in a **wind tunnel**. The smoke allows the designers to see how well the air moves over the shape.

b What do you notice about the airflow round these 2 cars?

▶ A racing cyclist needs to find ways to reduce air resistance as much as possible.

c Why is this important?

d List all the ways that you can see in the photo.

▶ This lorry has a wind deflector fitted above the cab.

e Explain carefully why this saves money.

Moving through water

In the same way, as objects pass through water there is a friction force that slows them down.

f How does the shape of a dolphin help it to move easily through the water**?**

▶ Olympic swimmers try to reduce friction between themselves and the water.
They wear tight smooth costumes, and swimming caps.

g Some swimmers shave all the hair from their bodies. Why does this improve their times**?**

▶ The diagram shows a model boat, with 4 forces on it:

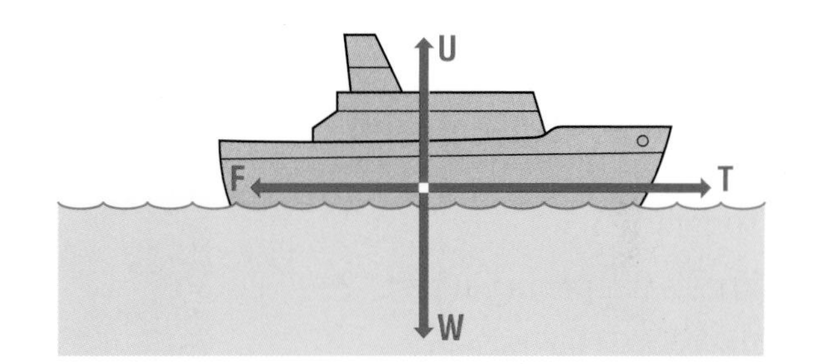

W, the **W**eight is 5 newtons.
The water pushes up with the force **U**, the **U**pthrust.
U and **W** are **balanced forces**.

h What is the size of force **U?**

T is the **T**hrust force of the engine.

i What is the name of the force **F?**

In the diagram, the size of **T** is 10 N, while **F** is 6 N.

j What is the **resultant force** on the boat**?**

k Which way does the boat move**?**

l What happens to the speed of the boat**?**

m What happens to the size of the force **F** as the boat moves faster**?**

n What happens when **F** and **T** become balanced forces**?**

Petrol consumption

The table shows some data for a car travelling on a level motorway. Look at the table carefully:

o Plot a graph of the data.

p Explain what the graph tells you. Can you explain its shape**?**

speed (mph)	20	30	40	50	60	70
(km/h)	32	48	64	80	97	113
petrol used (cm^3/km)	60	65	73	86	110	156

Things to do

1 Copy and complete:
a) When an object moves through air, there is a friction force called air . . . . (or . . . .).
b) This friction is reduced when the shape of the object is . . . .
c) When the friction between a car and the air is reduced, it can go . . . .

2 Explain the following:
a) How is the friction between a boat and water reduced in a hovercraft?
b) How does a parachute help a skydiver to descend safely?

3 Draw a design for a streamlined futuristic car. Remember that it has to be comfortable to sit in.

4 When the space shuttle (or a meteorite) enters the Earth's atmosphere, it glows red-hot.

Explain why. Use the particle theory if you can.

Speed on graphs

Learn about:
- distance–time graphs
- speed–time graphs
- acceleration

Tanni Grey winning a Gold Medal at the Paralympic Games

a The athlete in the photo can travel 70 metres in 10 seconds. What is her average speed?

b What does average mean? Does it mean she travelled at exactly this speed for all 10 seconds?

To see how speed varies during a journey, it helps to draw a graph. It can show you where you speed up or slow down.

There are 2 kinds of graph you can use. You used a distance–time graph in lesson 7K6 in Book 7.

1. Distance – time graph

Here is a **distance**–time graph for a sprinter:

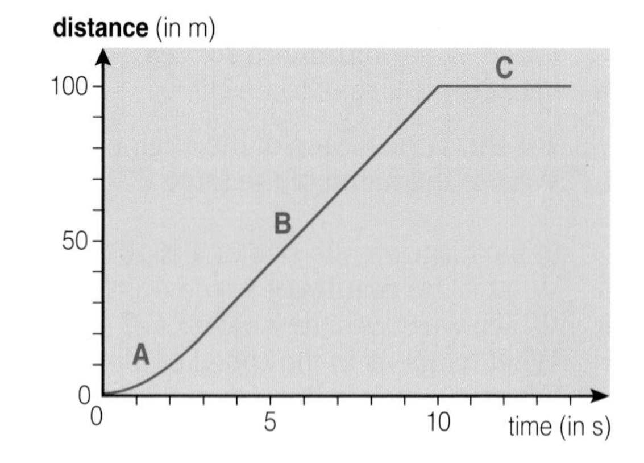

c What is happening in part C?

d How long is the race in (i) distance? (ii) time?

e What was her average speed?

f Why is the graph curved at A?

2. Speed – time graph

This graph shows the **speed**–time graph for a **car journey**. The graph has 4 parts, labelled **P**, **Q**, **R**, **S**. Look at the labels on the axes:

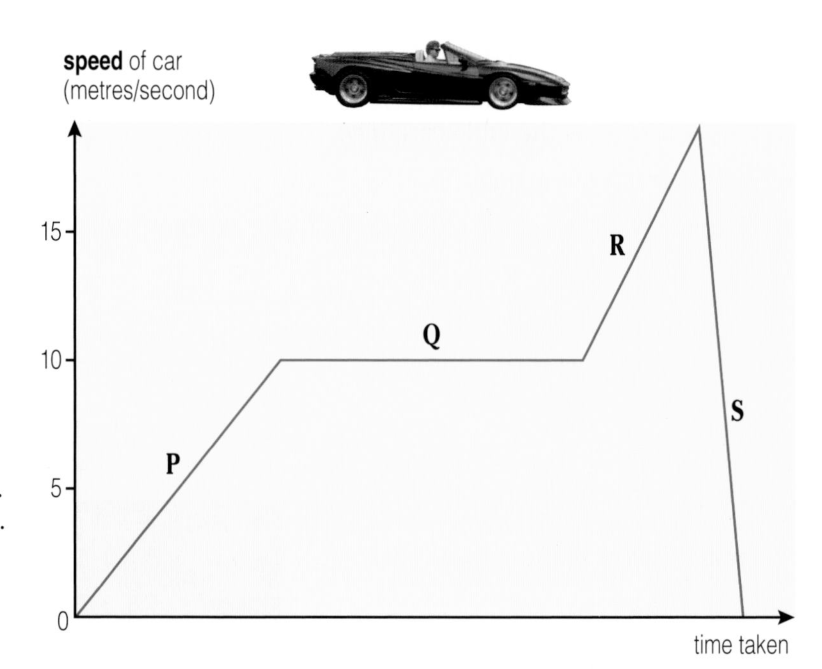

Part **P**.
Here the car is speeding up (***accelerating***). As time passes (along the graph), the speed increases (up the graph).

Part **Q**.
It is travelling at a ***constant speed*** of 10 m/s. The forces on the car must be balanced.

Part **R**.
The car ***accelerates*** to travel at a higher speed. There must be a resultant force, acting forwards.

Part **S**.
The speed of the car drops very quickly to zero. Perhaps it has crashed into a wall!

g Jack starts his car, accelerates to 10 m/s, keeps at this speed for a while and then brakes quickly to a stop. Sketch the speed–time graph for his journey.

Parachutes

Here is a speed–time graph for a **sky-diver** who jumps out of an aeroplane:

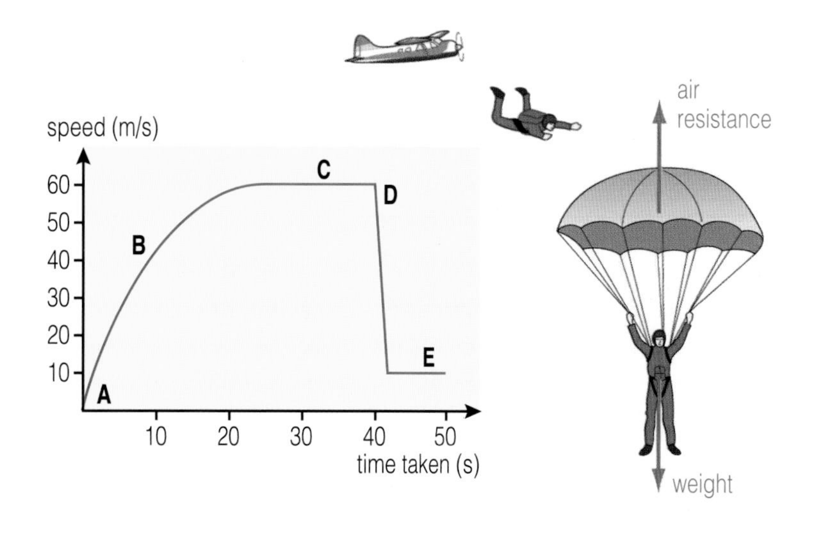

h At which point (A, B, C, D, E) did he jump?

i In which part does he speed up (accelerate) due to the pull of gravity (his weight)?

j After a while, the air resistance balances his weight, and so he stops going any faster. What is his speed now?

k Which point shows his parachute opening?

l What is his speed with the parachute open?

m Why is he slower than before?

Bungee-jumping

Look at this woman jumping off a bridge:

n What is the main force on her?
o Which way is it acting?

The speed–time graph below shows how she falls.

Point A was when she jumped off the bridge. She is fastened to an elastic rope.

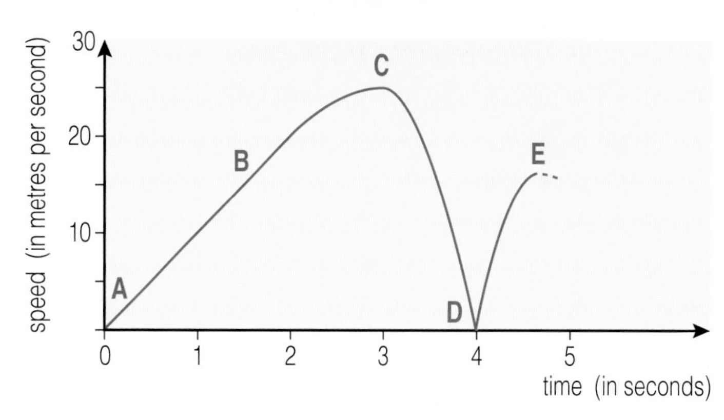

p At which point (B, C, D, E) is she moving fastest?
q What is her highest speed?
r At which point has she got most kinetic energy?

s At which point did she have most gravitational potential energy?

t Is she moving at point D?
u At which point is the rope stretched the most?

v Can you predict the shape of the rest of the graph?

Things to do

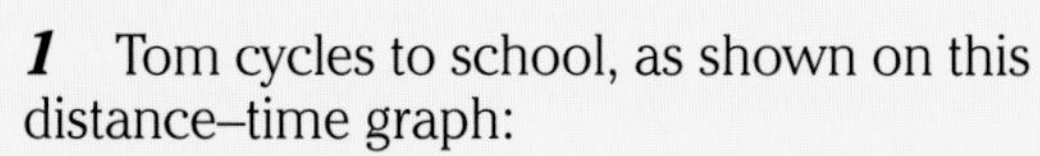

1 Tom cycles to school, as shown on this distance–time graph:

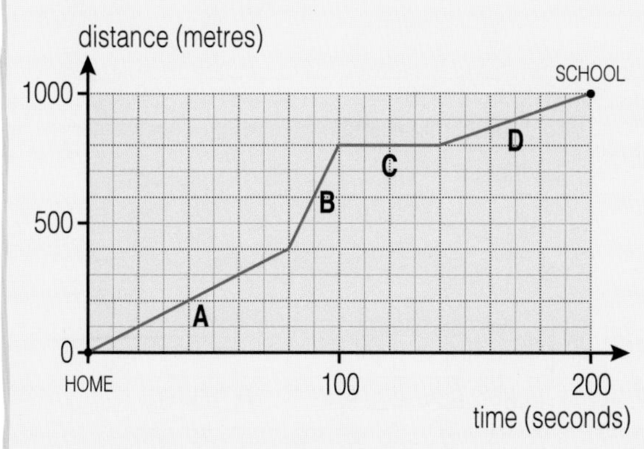

a) How far is it from home to school?
b) Which parts show that Tom is moving?
c) For how long did he stop at some traffic lights?
d) Which part is most likely to show him going down a hill? Explain why.
e) How far did he travel during part A?
f) Calculate his speed during part A.
g) Calculate his speed during part B.
h) Calculate his average speed for the whole journey.

2 Sketch a **speed**–time graph for Tom's journey. Add as much detail as you can.

Questions

1 There are 2 forces acting on this parachutist.
a) What is the name of the force acting upwards?
b) Name the force acting downwards.
c) If the man weighs 600 N, and air resistance is only 500 N, what is the resultant force?
d) What happens in this case?
e) What happens when his parachute opens?
f) Can you explain this, using the particle theory?

2 Fudge the cat goes for a little walk.
Here is her distance–time graph:

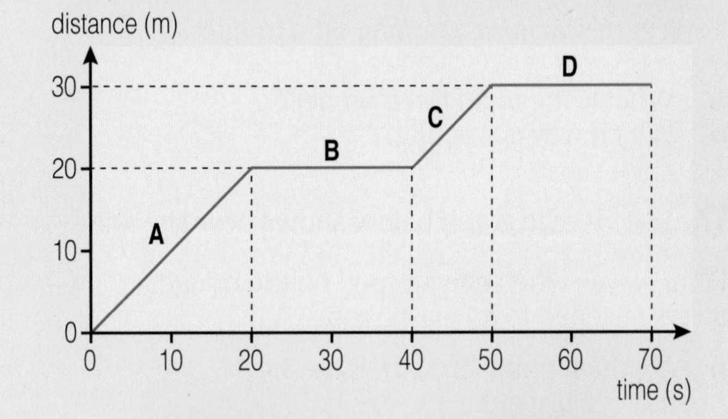

a) Use the scales on the graph to describe her journey in as much detail as you can.
b) What is her speed in part A?
c) What is her speed in part C?
d) Sketch the graph you would get if she runs twice as fast in parts A and C.
Add the correct numbers to your axes.

3 Calum makes a journey on his bike.
Sketch a distance–time graph that shows his journey:
He sets off from home, riding slowly for 1 minute.
Then he stops at the traffic lights for 1 minute.
He sets off again, riding at exactly the same speed as before, for 2 minutes. Then he stops for a rest for 1 minute.
Finally he rides very quickly down a hill for 1 minute.

4 Describe, in as much detail as you can, the motion of a car which has this speed–time graph:

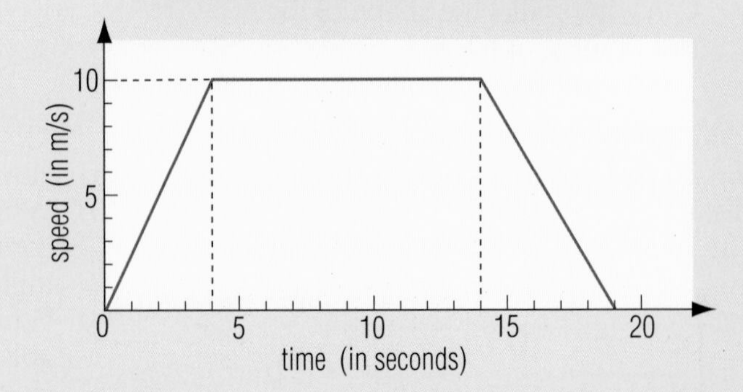

5 A car drives off from rest and travels in a straight line.
The speedometer readings at different times are shown in the table:

a) Plot a graph of this data.
b) Describe in words what happens during this journey.

time (s)	0	10	20	30	40	50	60	70	80	90	95
speed (m/s)	0.0	2.1	6.9	17.8	22.6	24.4	22.9	16.7	10.0	3.3	0.0

Pressure and moments

Your life is full of forces. You use forces to move around and to transfer energy.

In this unit we look at how forces can exert a pressure, and see how pressure can act on solids, liquids and gases.

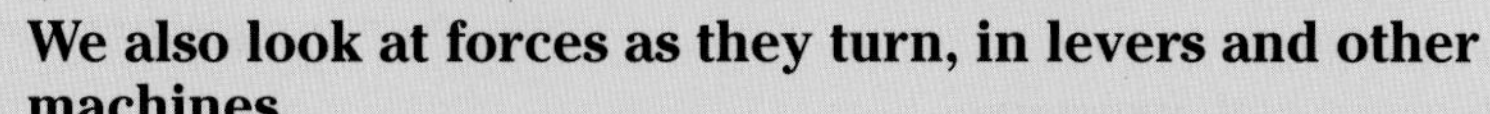

We also look at forces as they turn, in levers and other machines.

In this unit:

Under pressure

Learn about:
- the effects of pressure
- calculating pressure

▶ You can push a drawing-pin into the table:

But you can't push your thumb into the table, even though you use a bigger force.

Why do you think this is**?**

force

force

small area

large area

▶ The pin-point has a small area. All the force is concentrated in that area, to give a high **pressure**.

With your thumb, the same force is spread out over a larger area. The pressure is smaller.

a What is the real difference between a sharp knife and a blunt knife**?**

▶ Here is another example:

Why does the boy sink into the snow, while the skier stays on top**?**

What are snow-shoes**?** Why do eskimoes wear them**?**

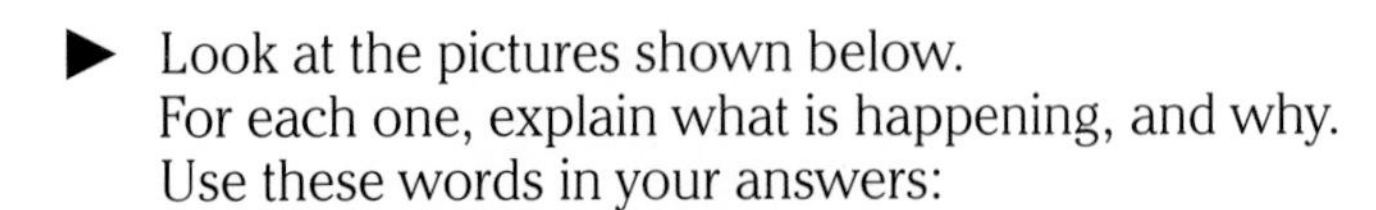

▶ Look at the pictures shown below.
For each one, explain what is happening, and why.
Use these words in your answers:

high/low **pressure** **force** small/large **area**

To calculate the pressure you need to know 2 things:
- the force exerted (in newtons)
- the area it is spread over (in cm^2 or m^2).

In fact:

$$\textbf{Pressure} = \frac{\textbf{force}\ \text{(in newtons)}}{\textbf{area}\ (\text{in } cm^2 \text{ or } m^2)}$$

If the area is in cm^2, then the unit of pressure is newtons per square centimetre (**N/cm^2**).

If the area is in m^2, then the unit of pressure is newtons per square metre (**N/m^2**).
This unit is also called a **pascal** (**Pa**). 1 Pa = 1 N/m^2.

Example

Nellie the elephant weighs 40 000 newtons.

She stands on one foot, of area 1000 cm^2.

What is the pressure on the ground?

Answer

$$\textbf{\textit{Pressure}} = \frac{\textbf{force}}{\textbf{area}}$$

$$= \frac{40\ 000\ \text{N}}{1000\ \text{cm}^2}$$

$$= \underline{40\ \text{N/cm}^2}$$

b Use the same method to calculate the pressure when a woman weighing 500 N stands on a stiletto heel of area 1 cm^2.

c Now compare these two pressures.
Which exerts the bigger pressure – the elephant or the shoe?
Which exerts the bigger force?

Shoe pressure

Plan a way to work out the pressure that you exert on the ground.

- Which 2 things do you have to measure?
- How can you do this?

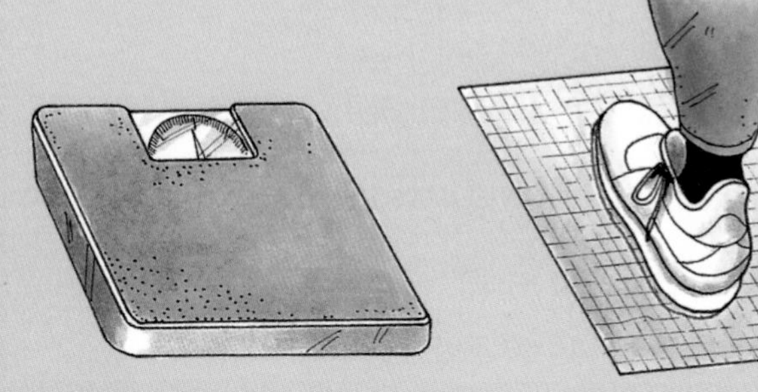

- If you have time, do the investigation.
- Who makes the biggest and least pressures in your class?

Things to do

1 Copy and complete:
a) The formula for pressure is:
b) Its units are . . . . per square centimetre (N/cm^2) or . . . . per square metre (N/m^2, also called . . . .).

2 Explain the following:
a) It hurts to hold a heavy parcel by the string.
b) It is more comfortable to sit on a bed than on a fence.
c) Heavy lorries may have 8 rear wheels.

3 Design a beach-chair suitable for use on soft sand. Sketch your design.

4 What is the pressure when a force of 12 N pushes on an area of 2 square metres?

5 A man weighs 800 N. The total area of his 2 shoes is 400 cm^2.
a) What is the pressure on the ground?
b) He puts on some snow-shoes of total area 1600 cm^2. What is the pressure now?

6 This box weighs 100 N.
a) Calculate the area of each face.
b) Calculate the pressures when i) the red, ii) the yellow, and iii) the blue faces are on the ground.

5 cm
10 cm
2 cm

Pressure all around

Learn about:
- pressure in liquids
- pressure in gases

Pressure in liquids

▶ Why do you think deep-sea divers have to wear special clothes? Why do submarines have to be strong?

Yes – it is because of the **pressure** on them.

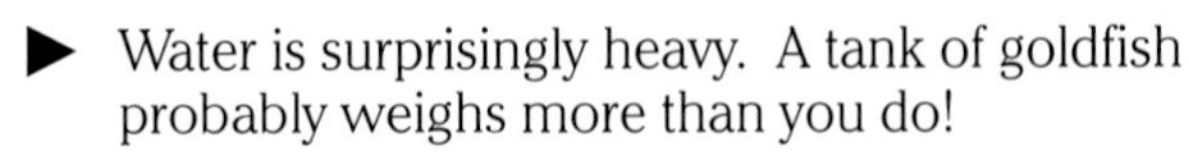

▶ Water is surprisingly heavy. A tank of goldfish probably weighs more than you do!

The weight of water presses down and exerts a pressure. As a fish swims deeper, the pressure on it increases more and more.

This tank has 3 holes in the side:

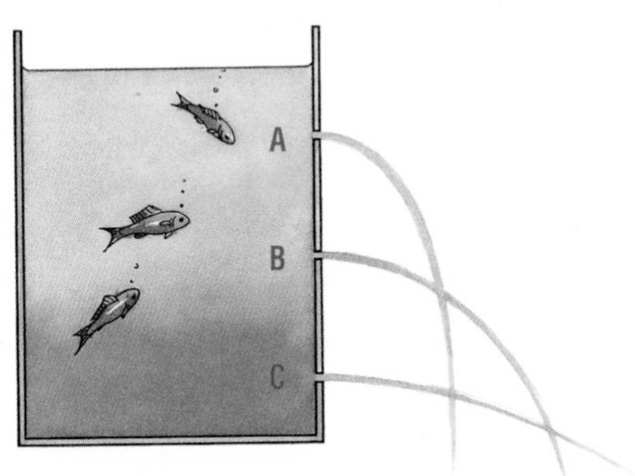

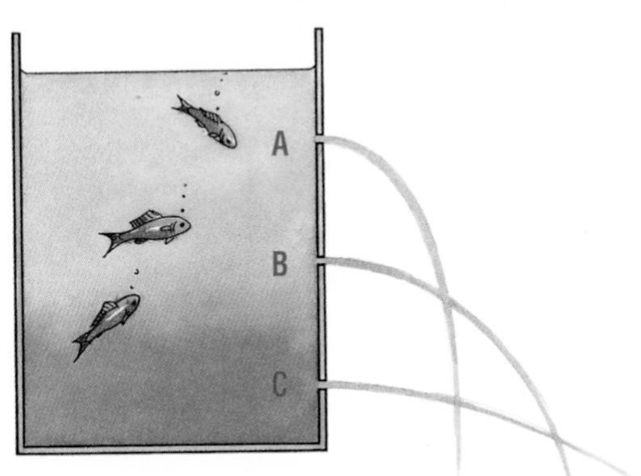

You can see that the water is spurting out under pressure.

Which jet of water is spurting out farthest?

Where is the pressure greatest – at A, B, or C?

▶ Here is a side-view of a lake made by a dam:

The length of each arrow shows the size of the pressure.

a Where is the pressure least?

b Where is the pressure greatest?

c Why does the dam need to be thicker at the bottom than at the top?

d Which forces are holding up the boat?
These forces are called the **upthrust**.

e The weight of the boat is 1000 N. Where does this force act?

f The upthrust and the weight of the boat are **balanced** forces. What does this mean?

g How big is the upthrust in this case?

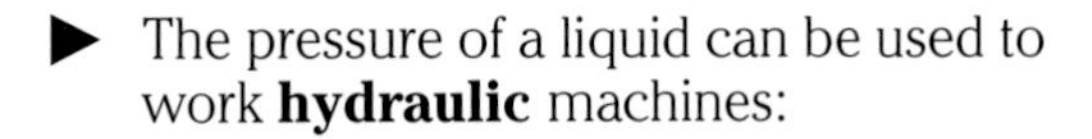

▶ The pressure of a liquid can be used to work **hydraulic** machines:

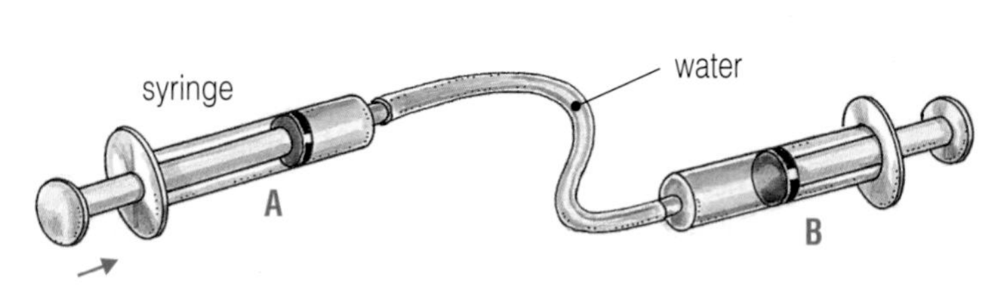

What happens if you push in piston A?

Where is this used on a car?
(Hint: see page 160.)

Pressure of air

Just as fish live at the bottom of a sea of water, we are living at the bottom of a 'sea' of air.
This air is called the **atmosphere**.

It exerts a pressure on us (just as the sea squeezes a fish).
This is called **atmospheric pressure**.

▶ If you blow up a balloon, you blow millions of tiny air particles (molecules) into it:

These molecules bounce around inside the balloon. Whenever a molecule hits the balloon, it gives the rubber a tiny push. Millions of these tiny pushes add up, to make the air pressure.

You are being punched on the nose billions of times each second by these tiny air molecules!

Can it be crushed?

Your teacher will show you an experiment with a metal can:

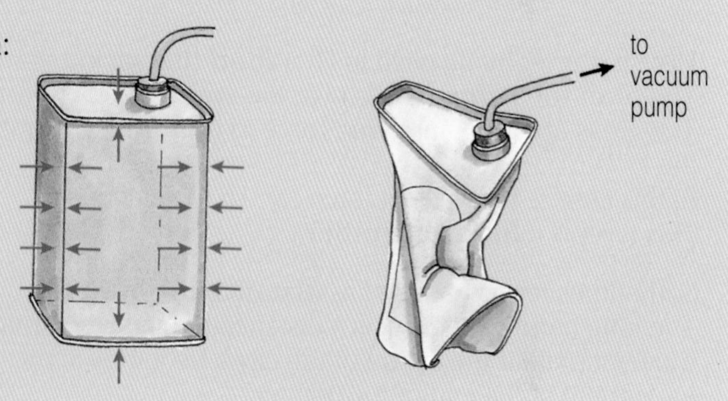

- At the start, the can has air inside and outside. The air molecules are bouncing on the inside and the outside.
- If a pump is used to take the air out of the can, what happens?
- Can you explain this? Try to use these words:
  molecules **air pressure**

Sucking a straw

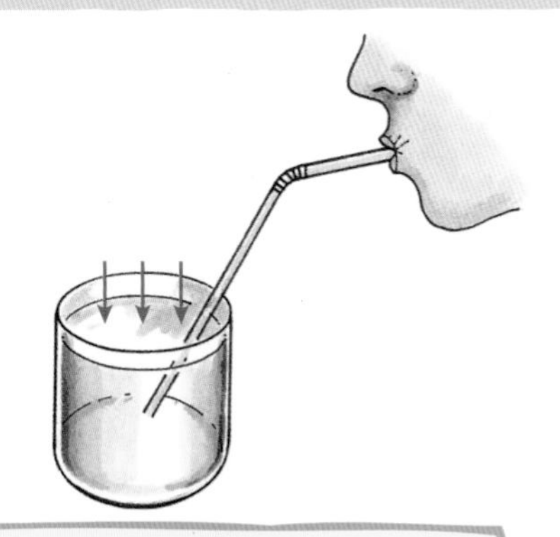

When you 'suck' on a straw, you use your lungs to lower the pressure inside the straw.

Explain why the liquid moves up, using these words:
molecules **air pressure**

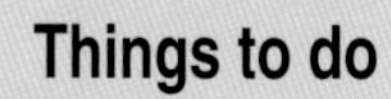

1 Copy and complete:
a) The pressure in a liquid is . . . . at the bottom than at the top.
b) A boat floats because the water pressure makes an . . . . force on it.
c) Air pressure is caused by billions of tiny . . . . bouncing around.

2 What happens to the air pressure as you go up a mountain? Why is this?

3 Explain why:
a) You can fill a bucket from a downstairs tap faster than from an upstairs tap.
b) Aeroplanes are often 'pressurised'.
c) Astronauts wear space-suits.

4 Alex says, "A vacuum cleaner works rather like a drinking straw."
Explain what you think he means, using the words: molecule, pressure.

Forces at work

Learn about:
- levers as force-magnifiers and distance-magnifiers
- pulleys

Levers

A lever is a simple **machine**. It helps us to do jobs more easily.

▶ The diagram shows two spanners, used for turning a nut on a bolt.

Which spanner would you use to undo a very tight nut? How can you make it even easier to undo the nut?

▶ Look at the door handle in the diagram: It is a lever. Which is the best place (A, B or C) for you to apply a force? (Try this if you can.)

▶ Design a simple machine or tool that would help an old person to turn a stiff door handle. Draw a diagram and label it.

Here is a lever being used to lift a ***load*** (the sack):
The pivot (or '***fulcrum***') is near the sack.
The girl is applying an ***effort*** force to turn the lever and lift the load.
In this case, a small effort force can lift a large load.
This lever is a **force-magnifier**.

Sometimes levers are used as **distance-magnifiers**.
The long hand on a clock is a distance-magnifier; a small movement near the centre of the clock becomes a big movement at the end of the pointer.
There are levers in your body that are used like this.

a Describe how your arm muscles work (see 9B7, page 30).

Here are some common machines using levers.
For each one, write down:

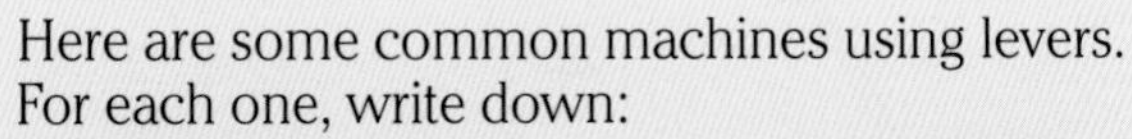

- where the pivot is
- where the effort force is applied (by you)
- where the load is
- whether it is a force-magnifier or a distance-magnifier.

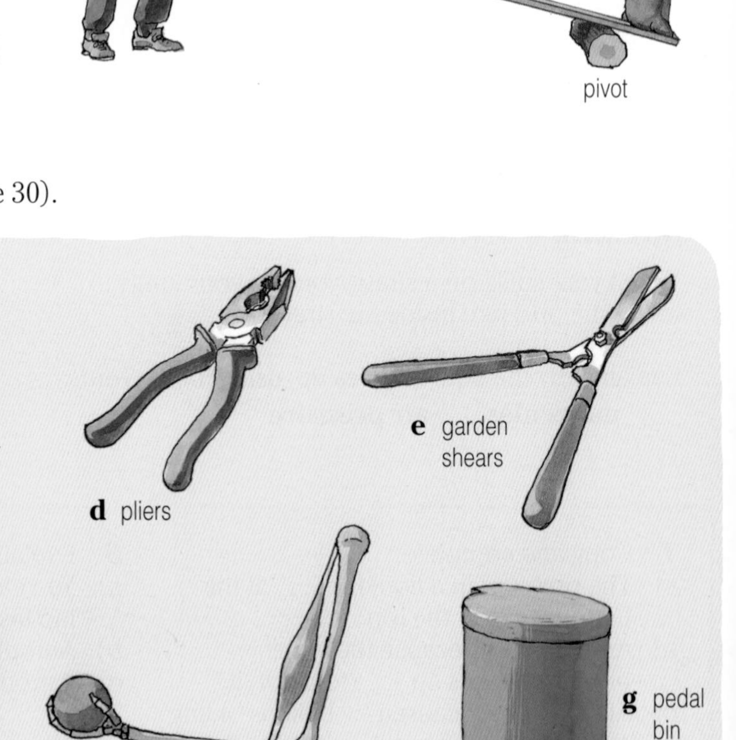

Pulleys

A **pulley** is another simple machine.

Here are two pulley systems that could be used on a building site:

What are pulley systems useful for**?**

Investigate these two pulley systems.

You can use a clamp stand to hold up the top pulley. You can use slotted weights for the load. You can measure the effort force with a force-meter (spring-balance).

Measure the effort needed to lift up different loads.

What do you find**?**

Which pulley system is easier to use**?**

Ring a bell

The diagram shows a design for a door-bell in an old castle.

When the handle is pulled:

h What happens to the blue rope**?**

i Which way does pulley A turn**?**

j Which way does lever B move**?**

k Which way does the red rope move**?**

l What happens to the bell**?**

m Is lever C a force-magnifier or a distance-magnifier**?**

n Can you design a simpler system**?** Draw a diagram of it.

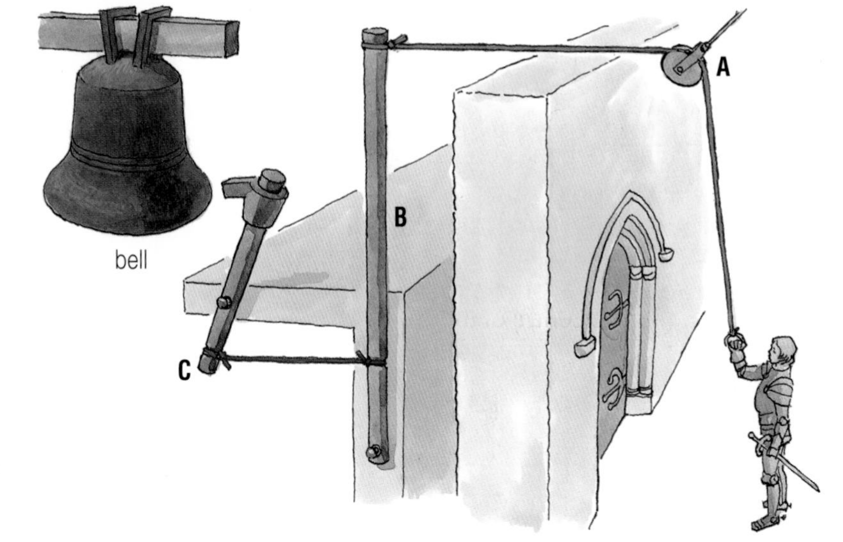

1 Copy and complete:
a) Levers and pulleys are simple . . . .
b) A lever has a pivot or . . . .
c) A lever can be a . . . . -magnifier or a . . . . -magnifier.

2 Think about the ways you can move your arms, fingers, legs, jaw, etc.
Make a list of the levers in your body.
For 3 of these levers, draw a simple diagram and show where the pivot is.

3 Design a gadget to help an invalid or an old person open a screw-top bottle.

4 Imagine you are using a spanner to undo a tight nut. Copy and complete this Energy Transfer Diagram (see page 110).

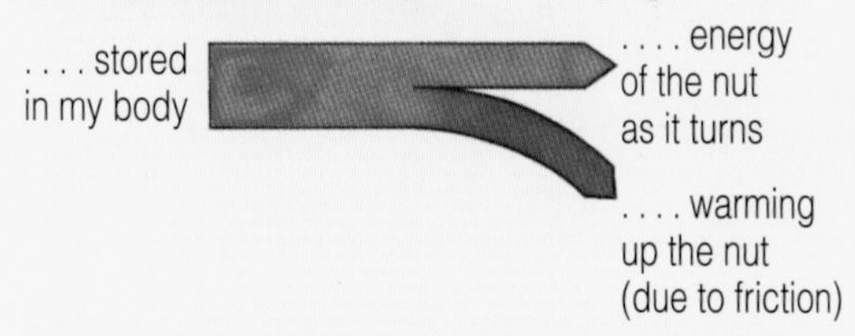

5 Design a system of pulleys and levers that will allow you to open the front door of your house from your bedroom.
Sketch a diagram of your system.

Just a moment

Learn about:
- moments
- calculating turning effects
- the principle of moments

▶ Which is the best place to push on a door to open it – at the hinge or at the door edge? Why?

▶ Some water-taps are hard for old people or invalids to turn. Design a better tap for an old person.

▶ As you know, a longer spanner makes it easier to turn a nut. In fact, the turning effect depends on ***two*** things, as the next experiment shows.

Hold a ruler at the very end, and put an object on it (for example, a rubber):

- Put the object at different positions on the ruler. What do you notice?
- Try a heavier object, at different distances.

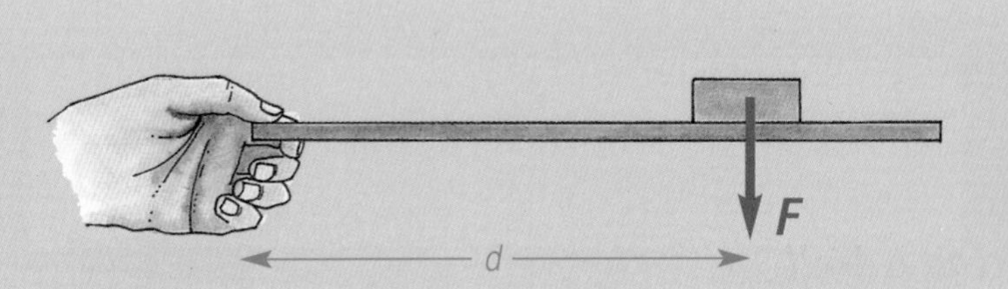

The turning effect depends on 2 things:
- the size of the force,
- the distance of the force from the pivot.

The turning effect of a force is called its **moment**.
The moment is calculated by:

Moment of the force = **force** (in newtons) × **distance from pivot** (in metres)

Moments are measured in units called **newton-metre** (**N m**).

Example 1

A car-driver is tightening a nut.

She exerts a 10 N force, 20 cm from the nut:

How big is the turning effect?

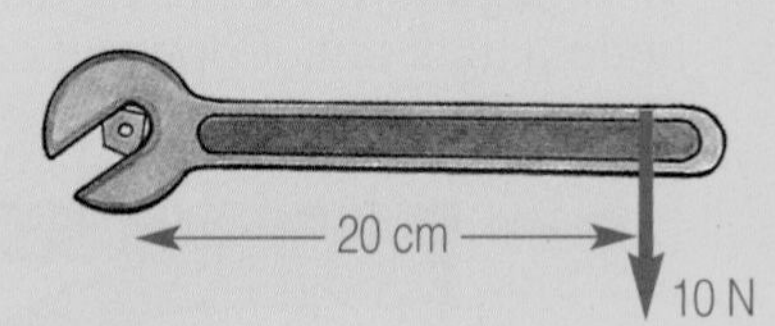

Answer

Distance from pivot = 20 cm = 0.2 m

Moment = force × distance from pivot
= 10 N × 0.2 m
= 2 N m

a If the driver applies a force of 100 N, 40 cm from the pivot, what is the moment?

b A boy pushes a door with a force of 10 N, 60 cm from the hinge. What is the moment?

Moments in balance

Here is a see-saw:

The big girl has a moment which is turning in a clockwise direction.

c Which way is the small boy's moment turning**?**

d Why do you think that the small boy can balance the big girl**?**

When the see-saw is balanced, and not moving, it is 'in equilibrium'.

Then: the **anti-clockwise moments** = the **clockwise moments**

This is called **the principle of moments** (or **the law of levers**).

Testing the principle of moments

Your teacher can give you a Help Sheet for this.

- You can use a ruler as a see-saw, and add weights:
- Work out the clockwise and anti-clockwise moments. What do you find**?**

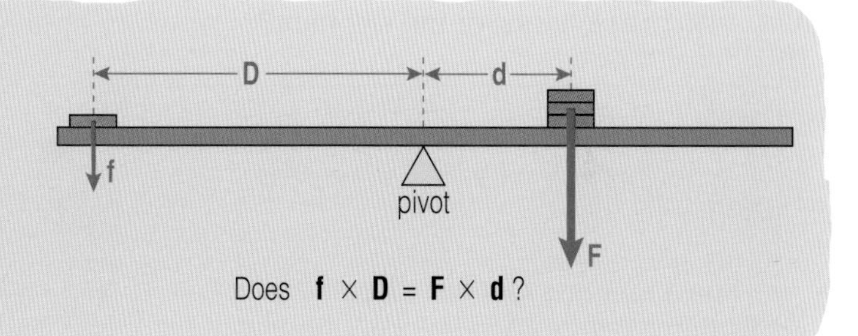

Example 2

A pole-vaulter is holding the pole:

His left hand acts as a pivot.
You can assume that the weight of the pole (50 N) acts at the centre of the pole, 1 m from the pivot.
How hard must his right hand push down**?**

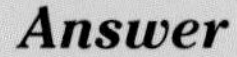

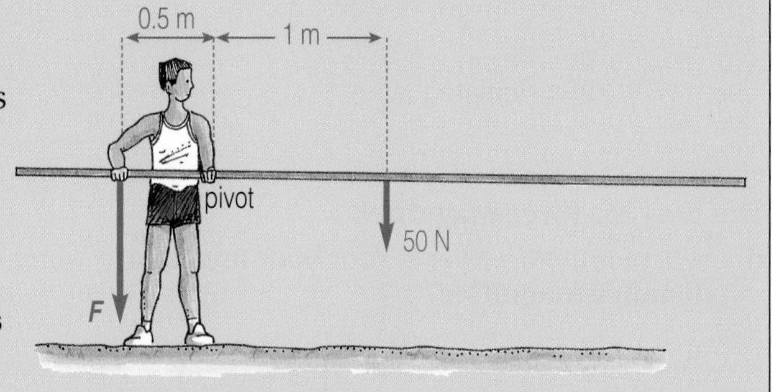

Answer

When balanced ('in equilibrium'):

anti-clockwise moments = clockwise moments

$$F \times 0.5 = 50 \times 1$$

$$F = \frac{50}{0.5} = \underline{100 \text{ newtons}}$$

Things to do

1 Copy and complete:

a) The turning effect (or . . . .) of a force is equal to the force multiplied by the . . . . from the . . . . Its unit is . . . .

b) The principle of moments states that, when an object is balanced and not moving, the anti-clockwise . . . . are . . . . to the . . . . moments.

2 Explain why it is difficult to steer a bike by gripping the centre of the handle-bars.

3 The diagrams show metre rules balanced at their centres.
What is the weight of a) *X*? and b) *Y*?

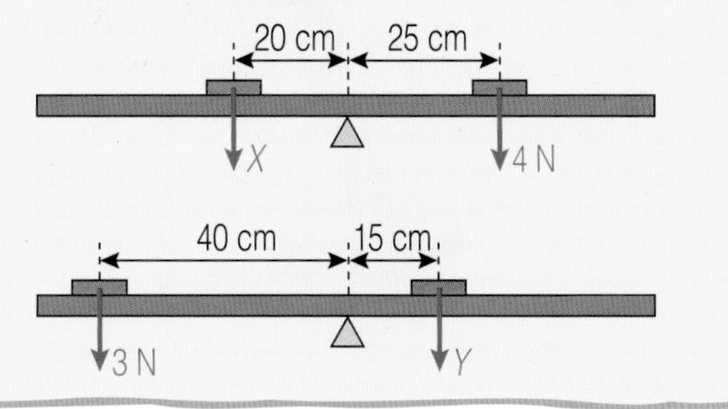

More about levers

Learn about:
- calculating effort
- centre of gravity
- stability

As you know, a lever is a simple machine.
It helps us to do work more easily.

a How many levers can you see in the room**?**

Here is a lever being used to lift a big **load**:
The lever can turn about a pivot (or 'fulcrum').
The man is applying an **effort** force.

b Where would you move the pivot to make the effort even smaller**?**

c Which moves through a bigger distance – the man's hands or the elephant**?**

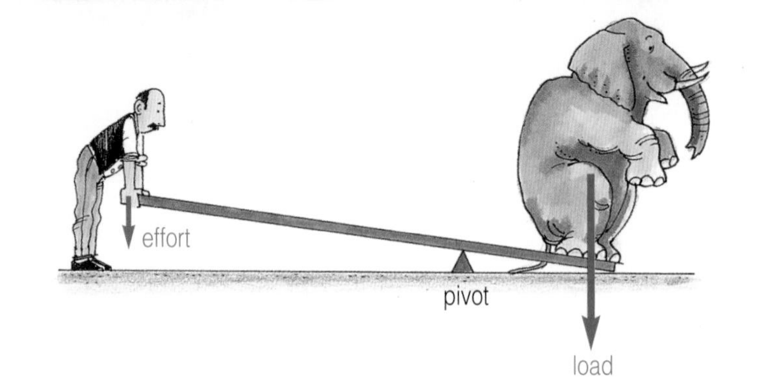

Example

A wheelbarrow with its load weighs 300 newtons.

The weight acts at a distance of 0.5 m from the centre of the wheel (the pivot).

What force is needed to lift the handles, which are 1.5 m from the centre of the wheel**?**

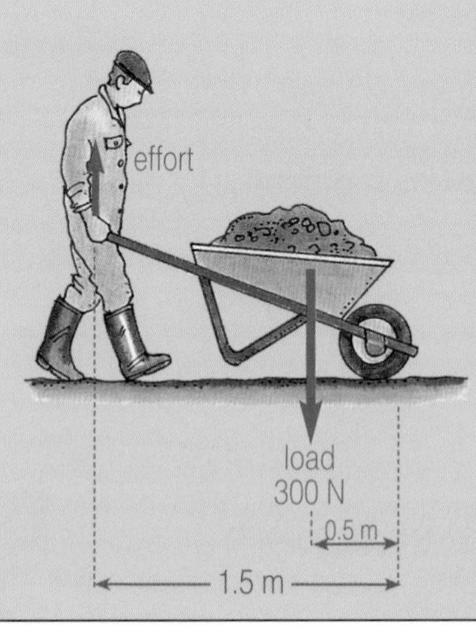

Answer

Use the principle of moments (p. 157).

When balanced ('in equilibrium'):

(clockwise moment = anti-clockwise moment)

effort × 1.5 = 300 × 0.5

effort = <u>100 newtons</u>

- Design a wheelbarrow that would need even less effort.

This lever is a **force-magnifier**.

d Where is there a lever in your body used as a **distance-magnifier?**

Here are some common machines using levers.
For each one you should be able to find the **pivot**, and decide the positions of the **effort** force and the **load** force.
Your teacher will give you a Help Sheet for this.

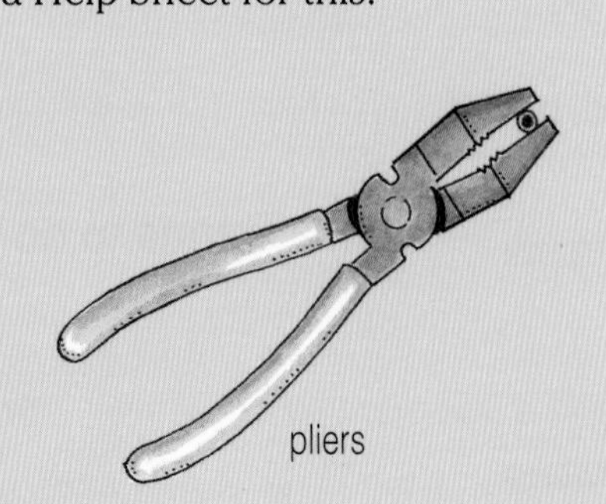

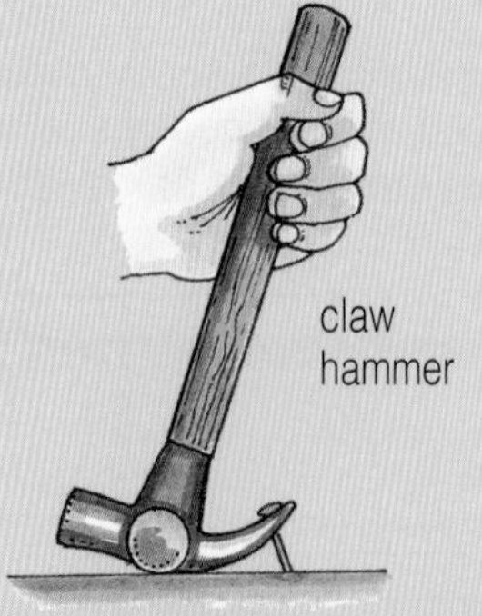

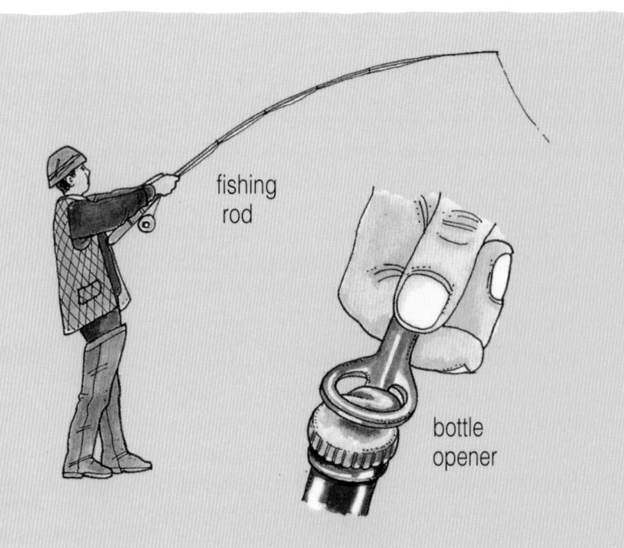

A balancing act

The gymnast is balanced on a beam:

Each part of her body is pulled down by gravity.

All the clockwise moments of the left-hand parts of her body are **balanced** by the anti-clockwise moments of the right-hand parts.

It's just as though all her weight is one force acting at one point **G**. G is called the **centre of gravity**.

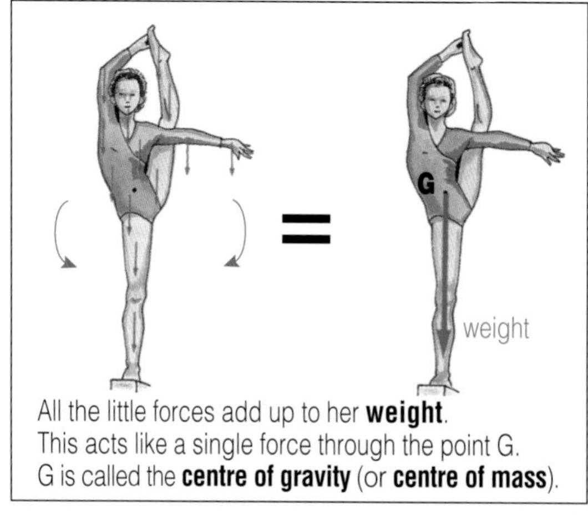

All the little forces add up to her **weight**. This acts like a single force through the point G. G is called the **centre of gravity** (or **centre of mass**).

e What would happen if her centre of gravity was not directly over her foot?

f Where is the centre of gravity of a metre rule?

Stability

If something is **stable**, it does not topple over. Use a box or match-box to investigate what makes an object stable.

- You can use plasticine to raise or lower the centre of gravity.
- To make the box more stable, should it have
  i) a high or a low centre of gravity?
  ii) a narrow or a wide base?
- Can you explain this, using the principle of moments?

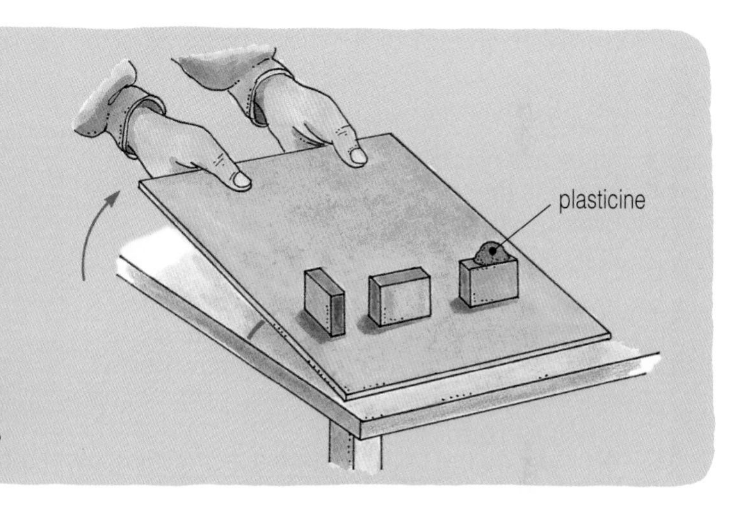

g Which is more stable: a racing-car or a double-decker bus?

h How are these 3 objects made stable?

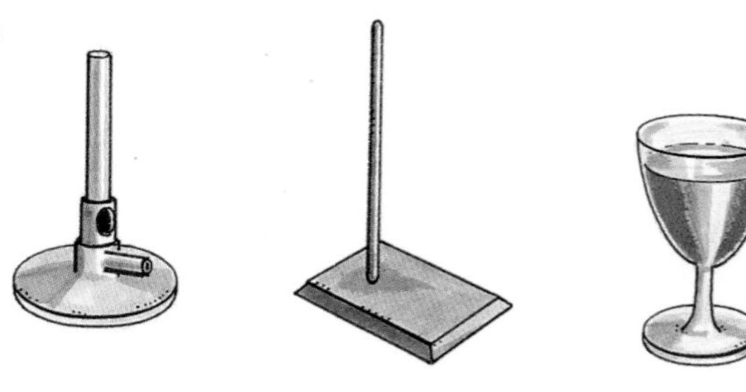

Things to do

1 Copy and complete:
a) A lever can be a . . . .-magnifier or a . . . .-magnifier.
b) The centre of . . . . (centre of mass) is the point through which the whole . . . . of the object seems to act.
c) A stable object should have a . . . . centre of . . . . , and a . . . . base.

2 Which is more stable: a milk-bottle
a) full? or b) $\frac{1}{4}$-full of milk?
Draw diagrams of the bottles on a slope to explain your answer.

3 The diagram shows a tall crane with a counter-weight to balance the load.
a) Calculate the size of the counter-weight.
b) What can you say about the position of the centre of gravity?

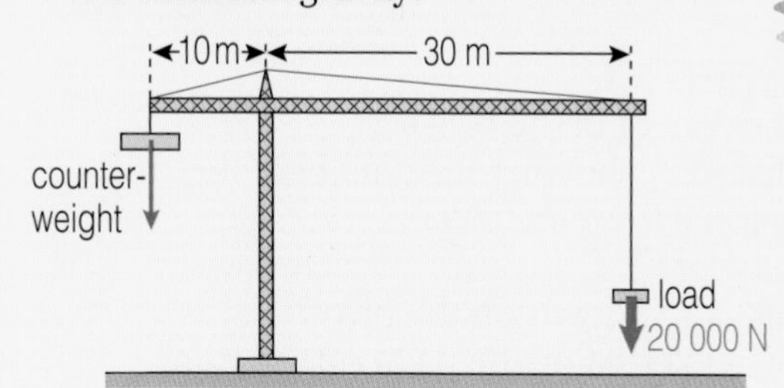

Questions

1 Jackie has a pair of stilts:

She weighs 400 N and the stilts weigh 100 N.
Each of her shoes has an area of 150 cm^2.
The bottom of each stilt has an area of 25 cm^2.
Calculate the pressure on the ground in each of the diagrams:

2 A car with 4 wheels has the tyres at a pressure of 20 N/cm^2.
Each tyre has an area of 100 cm^2 in contact with the road.

a) What is the weight of the car?
b) If 1 kg weighs 10 N, what is the mass of the car (in kg)?

3 Study this diagram showing the hydraulic brakes in a car:

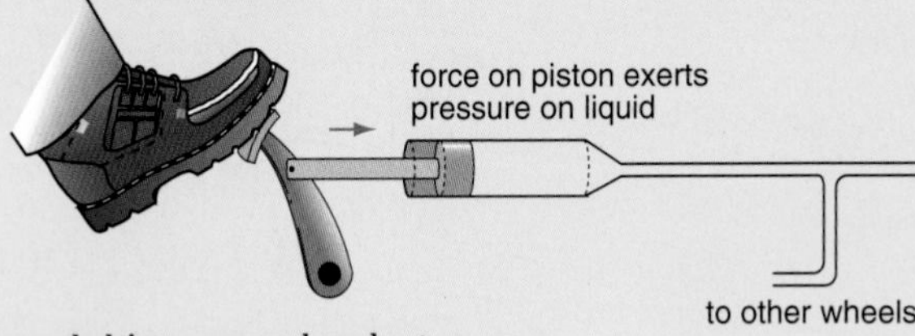

a) When the brake pedal is pressed, what happens to the liquid inside the tube?
b) Why does the rotating wheel slow down?
c) Where is friction (i) useful (ii) not useful in this brake system?
d) Why do the brake-pads in a racing-car glow red-hot?
e) Why does it help if the disc has a large area?

4 Melanie pumps up her bicycle tyre:
She notices that the pump becomes hot.

a) Where did the energy come from to pump up the tyre?
b) Explain how the air molecules exert a pressure on the walls of the tyre.
c) The air in the tyre was warmed up by the pumping. How does this affect the molecules of air?

5 The diagram shows a metre rule balanced on a pivot:

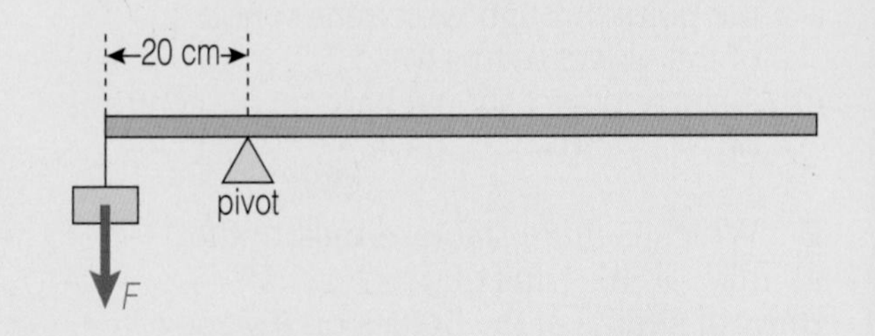

The weight of the rule is 1 N.

a) Where does the weight of the rule act?
b) Re-draw the diagram, showing the weight of the rule.
c) How far is this force from the pivot?
d) What is the value of *F*?
e) What is the total force down on the pivot?
f) What is the force exerted by the pivot on the rule?

Test tips

Sc
KEY STAGE 3
TIER 5–7

Science test

Paper 1

Please read this page, but do not open the booklet until your teacher tells you to start. Write your name and the name of your school in the spaces below.

First name ____________________

Last name ____________________

School ____________________

The national Tests usually take place in May in Year 9. As the date approaches, make sure that you are fully prepared.

Before the Tests:

- Plan your revision timetable.
  Your teacher can advise you on how much time you spend on each subject and when to start.
- Revise your work carefully.
  You can use the Revision Summaries on pages 162–173. These cover the main points from all your work in Years 7, 8 and 9.
- You can revise your Year 9 work by working through Question 1 in 'Things to do' at the end of each double page.
  Make sure you can fill in all the missing words.
  If you get stuck, ask your teacher for some help.
- Ask your teacher for copies of the Tests from earlier years.
  Try to do as many as you can, and time yourself.
- Make sure that you have the correct equipment needed for the Tests.
  This is usually:
  pen, pencil, rubber, ruler, protractor and calculator.
  A watch is also useful, so that you can pace yourself in each Test.

During each Test:

- Read each question carefully.
  Make sure that you understand what each question is about and what you are expected to do.
  For example, if it asks for ***one*** answer, give only ***one***.
- Sometimes you are given a list of alternatives to choose from.
  Make sure you choose the right number – if you include more answers than you are asked for, any wrong answers will cancel out your right ones!
- How much detail do you need to give**?**
  The question gives you clues:
  - Give short answers to questions which start: '***State …***' or '***List …***' or '***Name …***'.
  - Give longer answers if you are asked to '***Explain …***' or '***Describe …***' or if you are asked '***Why does … ?***'.
- The number of marks shown on the page usually tells you how many points the marker is looking for in your answer.
- The number of lines of space is also a guide to how much is needed in your answer.
- Don't explain something just because you know how to! If you're not asked to explain it, you won't gain any extra marks and you will just waste time.
- If a question is about a situation you haven't come across in class, don't panic. This type of question is seeing if you can apply your scientific knowledge to solve new problems.
- Some questions are designed to test your understanding of practical work and results from experiments. Study any tables or graphs carefully before answering.
- If you find a question too hard, go on to the next question. But try to write something for each part of every question.
- If you have spare time at the end, use it wisely to check over your answers.

Revision Summary: Biology 1

Health

- Being healthy involves having a balanced diet, taking suitable exercise, having a healthy lifestyle and being free from disease.
- Microbes include bacteria and viruses, some of which can cause disease.
- The body has natural barriers to infection, such as the skin.
- The production of antibodies and specialised cells in the blood are part of the body's defence system.
- Antibodies can protect you from disease by giving your body immunity.
- Antibodies can pass through the placenta to the fetus and through breast milk to the baby.
- Vaccines contain chemicals and cells which stimulate the body's defences.
- Antibiotics and other medicines can also help you fight disease.
- Not all diseases caused by microbes can be treated easily by drugs, e.g. some bacteria have become resistant to the effects of many antibiotics.
- Eating too much or too little of certain foods can affect your health.
- Exercise is important for healthy living. A person should take exercise that is appropriate to their age and body needs.

- Alcohol abuse can affect a person's health, lifestyle and family.
- Abuse of solvents and other drugs can affect your health.

Cells

- Animals and plants are made up of cells. Lots of cells grouped together make up a tissue. Different tissues make up organs, e.g. the heart is made up of muscle, nervous and connective tissues. Organs make up organisms, e.g. you have lungs, a heart, a brain, kidneys, a liver, etc.
- Cells are made up of: a membrane which keeps the cell together and controls what passes in and out of the cell; the cytoplasm (where the chemical reactions of the cell take place); and the nucleus (which controls the cell and contains instructions to make more cells).
- Plant cells have a thick cell wall to support the cell. Plant cells have chloroplasts to trap light energy for photosynthesis.
- Some cells have changed their shape to do different jobs. For example: Palisade cells in the leaf contain lots of chloroplasts to absorb light. Root hair cells are long and thin, with a large surface area to absorb water and nutrients. Cells lining your air passages have cilia (hairs) to move mucus up to your nose. Sperm cells have tails to swim to the egg and fertilise it. Egg cells have a food store to feed the fertilised egg.

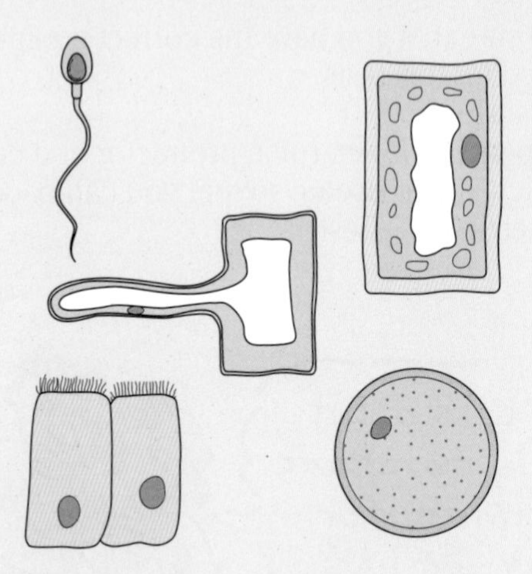

- Cells make new cells by dividing.
  When new cells are made, growth occurs and there is an increase in size.
- At fertilisation there is a joining together of male and female cells. In humans, a sperm joins with an egg. In flowering plants, a pollen cell joins with an ovule cell.
- These sex cells enable information to be transferred from one generation to the next, so we inherit characteristics from our parents.

Revision Summary: Biology 2

The active body

- Oxygen is used in your cells to release energy from food during respiration.
  oxygen + glucose → carbon dioxide + water + energy
- During respiration, glucose is broken down into carbon dioxide and water.
- The lungs, diaphragm and ribs and their muscles are all involved in breathing.
- Air gets to the alveoli of your lungs through the wind-pipe and air passages.
- Oxygen passes through the alveoli into the blood capillaries.
  Carbon dioxide passes the opposite way, from the blood capillaries into the alveoli.
- Tobacco smoke contains harmful chemicals including tar, nicotine and carbon monoxide that can damage your lungs and cause diseases such as lung cancer, bronchitis and heart disease.
- Your air passages have cells with cilia (hairs) which move mucus up to your nose. Mucus helps us by trapping dust and a number of microbes.
  Smoking prevents the cilia from working and so a person's lungs become congested with mucus. This causes 'smoker's cough'.
- Your heart needs to work efficiently in order to pump blood to the lungs and other parts of the body and return it to the heart.
- Your blood transports the oxygen, dissolved food, carbon dioxide and other waste chemicals around your body.
- Food and oxygen pass out of the blood capillaries into the cells.
  Carbon dioxide and other waste chemicals pass in the opposite direction.
- Your skeleton supports and protects your body and allows you to move.
- Your muscles provide the force needed to move bones at joints.
- When one muscle in a pair contracts, the other one relaxes.
  We say that they are antagonistic, e.g. your biceps and triceps muscles.

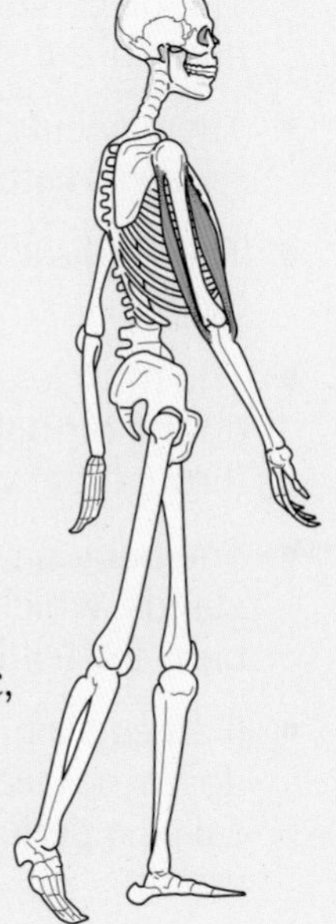

Food and digestion

- A healthy diet has a variety of foods, each in the right amount.
- A balanced diet contains carbohydrates, proteins, fats, minerals, vitamins, fibre and water.

- Different foods are rich sources of some of these, e.g. fish is rich in protein, and cereals have lots of carbohydrate and fibre.
- Carbohydrates and fats are used as fuel during respiration to release energy.
- We should balance our energy intake (the amount of energy in our food) with our energy output (the amount of energy our body uses up in a day).
- Proteins are needed for growth. We use them to make new cells and to repair damaged tissue.
- We need small amounts of vitamin A, vitamin B group, vitamin C, iron and calcium to stay healthy.
- We must digest our food before our bodies can use it.
  Digestion means breaking down large, insoluble molecules into small, soluble molecules.
- Enzymes can digest large food molecules, like starch, proteins and fats.
- The gut is a tube along which our food passes.
- Food has to be digested if it is to pass through the gut wall into the blood and transported to other parts of the body.
- Some material cannot be digested and is egested from the gut as waste.
- Malnutrition can involve either having too little or too much of certain foods.

Revision Summary: Biology 3

Variation and classification

- A species is a group of plants or animals that are able to reproduce fertile offspring (offspring that are able to breed).
- There is variation between *different* species, e.g. differences between dogs and cats.
- There is also variation between individuals of the *same* species, e.g. differences between domestic dogs.
- We inherit some features from our parents, e.g. eye colour.
- Other features are caused by our environment, e.g. a sun-tan.
- Some variation is *gradual*, e.g. height, whilst other variation is *clear-cut*, e.g. eye colour.

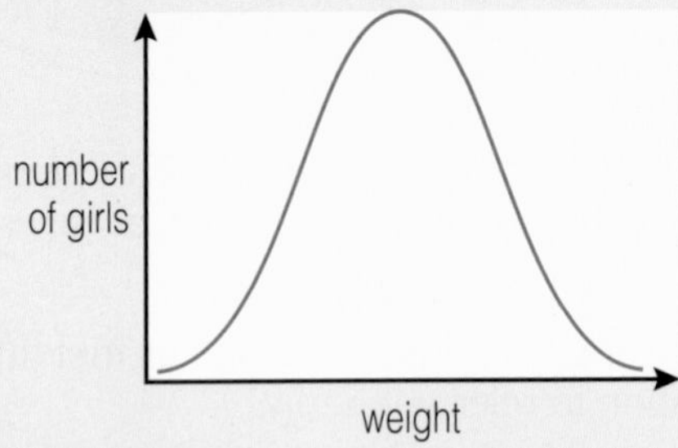

- Living things can be put into groups that have similar features.
- Animals with backbones are called vertebrates. They include fish, amphibians, reptiles, birds and mammals.
- Animals without backbones are called invertebrates. They include the jellyfish group, flatworms, roundworms, segmented worms, molluscs, arthropods (jointed-legged animals) and the starfish group.
- The main plant groups include the mosses, ferns, conifers and flowering plants.
- Some plants reproduce by making seeds, others by making spores.
- Useful features can often be bred *into* some animals and plants; less useful features can often be bred *out*. Selective breeding can produce new varieties of plants and animals.
- Scientists have been able to selectively breed high yield crops that are resistant to frost and resistant to disease.
- For years farmers have selectively bred cattle for their meat and milk, sheep for their wool, and pigs for their meat.

Reproduction

- At adolescence our bodies and our emotions change.
  A boy starts to make sperms, grow body hair and his voice deepens.
  A girl starts to make eggs, her breasts develop and she starts having periods.
- These changes are controlled by chemicals produced in our bodies called hormones.
- The sperm tube carries sperm to the penis; glands add fluid to make semen.
- The egg tube carries an egg to the uterus every month.
- The male sperm nucleus contains characteristics of the father and the female egg nucleus contains characteristics of the mother.
- During love-making, sperms are placed into the vagina.
- At fertilisation the sperm penetrates the egg and its nucleus joins with the egg nucleus.
- If the egg is fertilised it passes down the fallopian tube and settles into the uterus.
- The fertilised egg grows first into an embryo and then into a fetus.

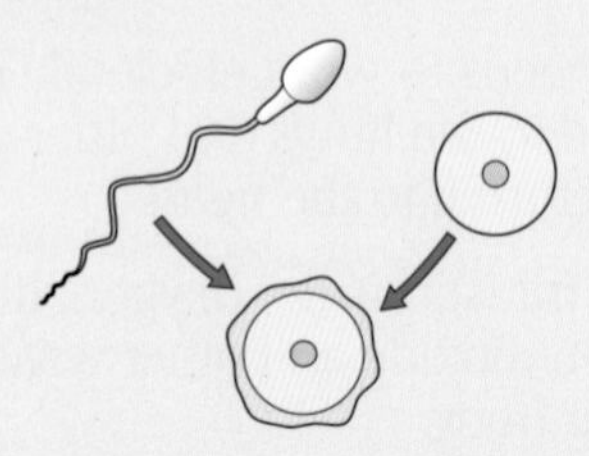

- The placenta acts as a barrier to infections and harmful substances.
- The placenta gives the fetus food and oxygen and removes carbon dioxide and waste chemicals.
- The fluid sac acts as a shock absorber to protect the fetus.
- The muscles in the wall of the uterus contract during birth, pushing out the fetus and placenta through the vagina.
- The baby is fed on milk from the mammary glands, which provide nutrients and protect the baby from infection.
- If an egg is not fertilised, the uterus lining breaks down and leaves through the vagina. This is called a period, which occurs about once a month.

Revision Summary: Biology 4

Ecology

- A habitat is a place where a plant or animal lives.
- Plants and animals can only survive in a habitat if they have all the things they need for life and reproduction.
  For example, plants need light, water, suitable temperature and nutrients if they are to survive.
- Different animals and plants are adapted to survive in different habitats.
  Many plants and animals are adapted to daily and seasonal changes in their habitats.
  Some plants are able to survive the winter as seeds.
  Some animals survive the winter by hibernating (going to sleep) or by migrating (flying to warmer countries).

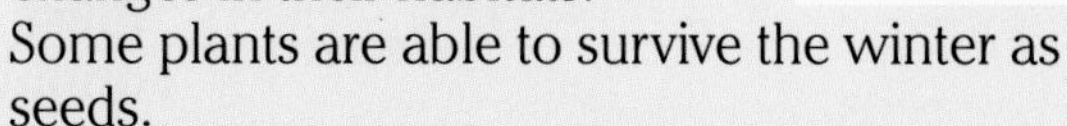

- A **population** is a group of animals or plants living in the same habitat.
- Some factors can limit the growth of a population, e.g. predators, disease, climate.
- Living things compete for resources that are in short supply, e.g. food, space.
  Living things that compete successfully survive to produce more offspring.
- Predators are adapted to kill other animals (their prey) for food.
  Many prey species have adaptations to avoid being caught, e.g. camouflage.
- **Food chains** show what an animal eats and how food energy is passed on.
- A food web is made up of many food chains.
- A pyramid of numbers shows the numbers of plants, herbivores and carnivores in a habitat.
- There is a flow of food energy from the producers to the final organisms in the food chain.

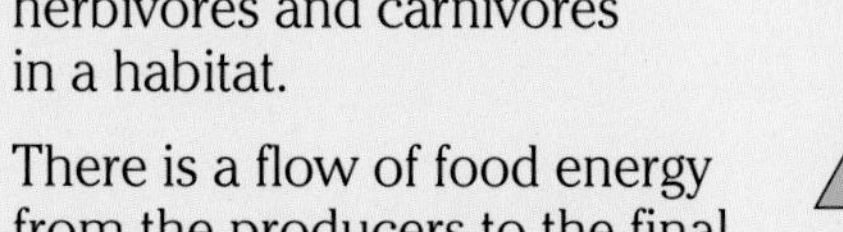

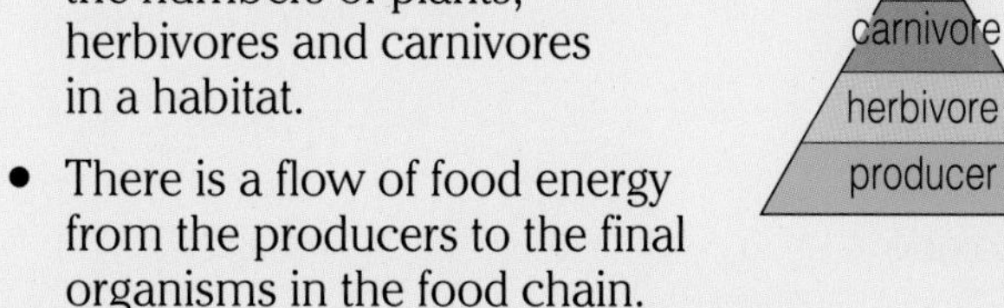

- Poisonous chemicals can increase in concentration along food chains.
- It is important to protect and conserve living things and their environment.
- Sustainable development involves not felling more trees or catching more fish than can be replaced by normal reproduction.

Plants at work

- Green plants use chlorophyll to trap light energy for photosynthesis.

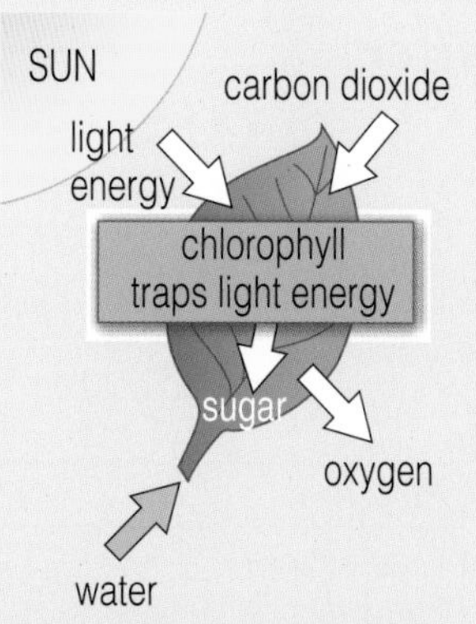

- During photosynthesis, green plants use this energy to convert carbon dioxide and water into sugar and oxygen:

$$\text{carbon dioxide} + \text{water} \xrightarrow[\text{chlorophyll}]{\text{light energy}} \text{sugar} + \text{oxygen}$$

- The new material that plants produce by photosynthesis is called biomass.
- Plants release oxygen for animal respiration and use up waste carbon dioxide.
- Plants also use oxygen for their own respiration.
- Plants also need nutrients from the soil, such as nitrates, for healthy growth.

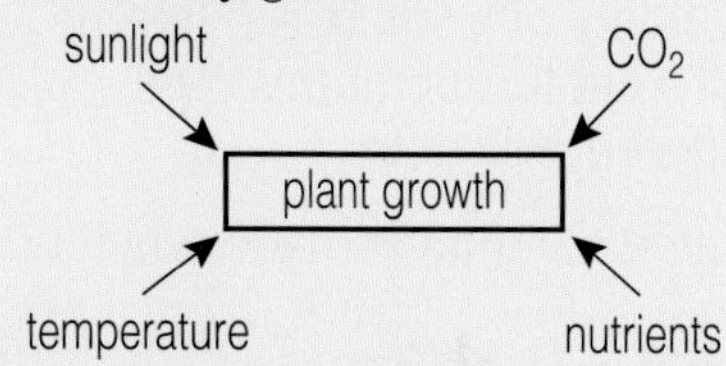

  Extra nutrients are sometimes added to the soil in the form of fertilisers.
- Root hairs absorb water and nutrients from the soil.
- Green plants are the producers that are the first link in a food chain.
- The rate of photosynthesis is affected by light, carbon dioxide and temperature.
- Green plants affect the amounts of carbon dioxide and oxygen in the atmosphere.
- Flowers are made up of sepals, petals, stamens and carpels.
- The anthers make pollen grains and the ovary makes ovules.
- Pollination is the transfer of pollen from the anthers to the stigma.
- Fertilisation happens when a pollen nucleus joins with an ovule nucleus.
- After fertilisation, the ovary changes into a fruit and the ovule grows into a seed.
  Under the right conditions the seed can grow into a new plant.

Revision Summary: Chemistry 1

Matter

- 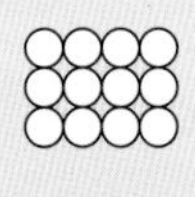 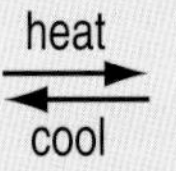

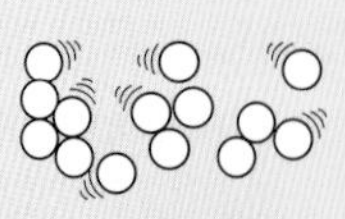 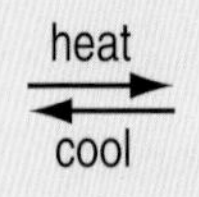

solid
Particles vibrate about a fixed position

liquid
Particles vibrate and change position

gas
Particles move freely in all directions

Gas particles hit the walls of the balloon. This causes gas pressure.

-

Properties	*Solid*	*Liquid*	*Gas*
Fixed shape?	Yes	No – shape of container	No – shape of container
Fixed volume?	Yes	Yes	No – fills the container
Easily squashed (compressed)?	Very difficult	Can be compressed	Easy to compress
Flows easily?	No	Yes	Yes
Dense (heavy for its size)?	Yes	Less dense than solids	Less dense than liquids

density = $\frac{\textbf{mass}}{\textbf{volume}}$

- 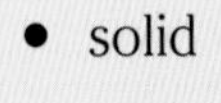 solid 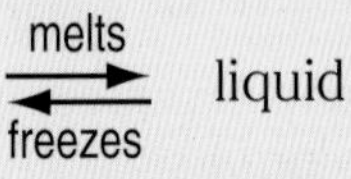

liquid 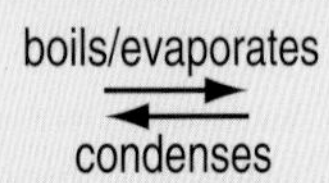

gas
- Different materials have different melting points, boiling points and densities.
- Most substances expand (get bigger) when they are heated.
  Most substances contract (get smaller) when they are cooled.
- Particles can move and mix themselves. This is diffusion.
- solute (soluble solid) + solvent (liquid) → solution
- The solubility of a solute is different ...
  – at different temperatures, and
  – in different solvents.

Expanding and contracting

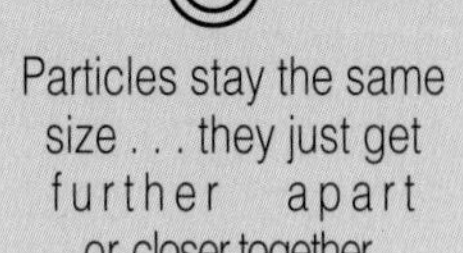

Particles stay the same size . . . they just get further apart or closer together.

- A saturated solution is made when no more solid can dissolve in solution at that temperature.

Acids and alkalis

- Acids and alkalis are chemical opposites.
  Alkalis are bases (metal oxides, hydroxides or carbonates) which dissolve in water.
- We can use indicators to show which things are acids and which things are alkalis.
  Coloured berries, flower petals and vegetables make good indicators.
- Universal indicator is a mixture of indicators. Its colour gives you the pH number of the substance.

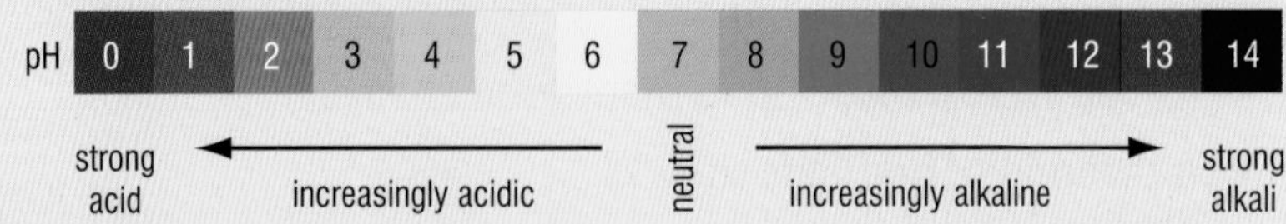

- Acids can be changed into salts by chemical reactions.
  acid + metal → a salt + hydrogen
  acid + base (alkali) → a salt + water
  e.g. acid + metal oxide → a salt + water
  e.g. acid + metal carbonate → a salt + water + carbon dioxide
  (carbon dioxide gas turns limewater cloudy)
- The acid + base reaction is called neutralisation.
  e.g. sulphuric acid (acid) + potassium hydroxide (base) → potassium sulphate (a salt) + water
  Neutralisations are useful reactions.
  e.g. Acids damage teeth. Toothpastes contain alkali to neutralise acids in the mouth.
  Some plants grow better in alkaline soil. Lime can be added to change the pH of the soil.
- Acids in the atmosphere can cause damage. Acid rain can corrode metal.
  It can dissolve rock (e.g. limestone).

Revision Summary: Chemistry 2

Elements

- Everything is made from very small particles called atoms.
  An element contains only one type of atom.
  Elements are simple substances. They cannot be broken down into anything simpler.
- Each element has a symbol, e.g. carbon is **C**, magnesium is **Mg**, iron is **Fe**.
- All the elements can be arranged in the periodic table.
  The columns of elements in the table are groups.
  The rows are periods.
- All the elements in a group have similar properties.
  They often react in the same way.
- There are 2 main types of element: metals and non-metals.
  Examples of metals are: copper, magnesium, tin, sodium.
  Examples of non-metals are: oxygen, hydrogen, carbon, sulphur.
- Metals are usually hard, shiny solids.
  They often have high melting points.
  They are good conductors of heat and good conductors of electricity.
  A few metals are magnetic, e.g. iron.
- Non-metals are usually gases or solids with low melting points. Most are poor conductors of heat and electricity. They are insulators.
  The solids are often brittle (they break easily).
- Metals are found on the left-hand side of the periodic table.
  Non-metals are found on the right-hand side.

Compounds and mixtures

- Each element contains only one type of atom.
  Compounds have 2 or more different atoms joined together.
- When atoms join together they make molecules.

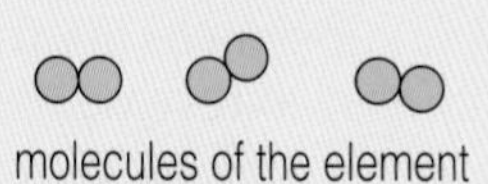

molecules of the element nitrogen, N_2

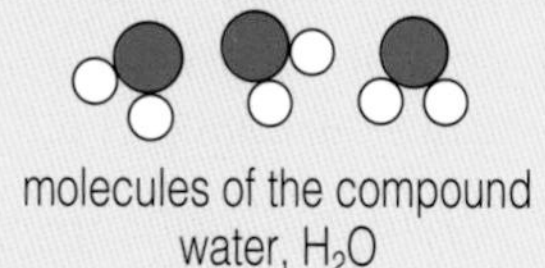

molecules of the compound water, H_2O

- Elements combine to make compounds.
  This happens in a chemical reaction.
  We can show a chemical reaction by a word equation:
  e.g. magnesium + oxygen → magnesium oxide
- A compound has a fixed composition.
  Each compound has its own formula,
  e.g. magnesium oxide is always MgO.
- A mixture contains more than one substance.
  It does not have a fixed composition.
  The parts that make up the mixture are not combined, e.g. air is a mixture of nitrogen, oxygen, carbon dioxide and other substances.
- Usually we can separate a mixture into pure substances. Methods we can use are:
  – filtration
  (to separate an insoluble solid from a liquid)
  – distillation
  (to separate pure liquids from solutions)
  – chromatography
  (to separate mixtures of colours).

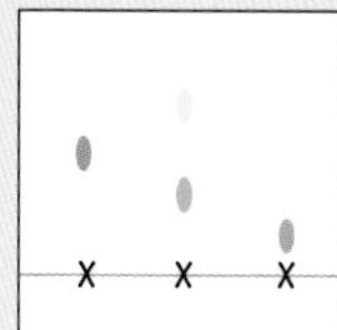

Key: Relative atomic mass / **Symbol** / Name / Atomic number

Period	I	II											III	IV	V	VI	VII	0
	s-block		d-block										p-block					
1	1.0 **H** Hydrogen 1																	4.0 **He** Helium 2
2	6.9 **Li** Lithium 3	9.0 **Be** Beryllium 4											10.8 **B** Boron 5	12.0 **C** Carbon 6	14.0 **N** Nitrogen 7	16.0 **O** Oxygen 8	19.0 **F** Fluorine 9	20.2 **Ne** Neon 10
3	23.0 **Na** Sodium 11	24.3 **Mg** Magnesium 12											27.0 **Al** Aluminium 13	28.1 **Si** Silicon 14	31.0 **P** Phosphorus 15	32.1 **S** Sulphur 16	35.5 **Cl** Chlorine 17	40.0 **Ar** Argon 18
4	39.1 **K** Potassium 19	40.1 **Ca** Calcium 20	45.0 **Sc** Scandium 21	47.9 **Ti** Titanium 22	50.9 **V** Vanadium 23	52.0 **Cr** Chromium 24	54.9 **Mn** Manganese 25	55.9 **Fe** Iron 26	58.9 **Co** Cobalt 27	58.7 **Ni** Nickel 28	63.5 **Cu** Copper 29	65.4 **Zn** Zinc 30	69.7 **Ga** Gallium 31	72.6 **Ge** Germanium 32	74.9 **As** Arsenic 33	79.0 **Se** Selenium 34	79.9 **Br** Bromine 35	83.8 **Kr** Krypton 36
5	85.5 **Rb** Rubidium 37	87.6 **Sr** Strontium 38	88.9 **Y** Yttrium 39	91.2 **Zr** Zirconium 40	92.9 **Nb** Niobium 41	95.9 **Mo** Molybdenum 42	99.0 **Tc** Technetium 43	101.1 **Ru** Ruthenium 44	102.9 **Rh** Rhodium 45	106.4 **Pd** Palladium 46	107.9 **Ag** Silver 47	112.4 **Cd** Cadmium 48	114.8 **In** Indium 49	118.7 **Sn** Tin 50	121.8 **Sb** Antimony 51	127.6 **Te** Tellurium 52	126.9 **I** Iodine 53	131.3 **Xe** Xenon 54
6	132.9 **Cs** Caesium 55	137.2 **Ba** Barium 56	138.9 **La** * Lanthanum 57	178.5 **Hf** Hafnium 72	180.9 **Ta** Tantalum 73	183.9 **W** Tungsten 74	186.2 **Re** Rhenium 75	190.2 **Os** Osmium 76	192.2 **Ir** Iridium 77	195.1 **Pt** Platinum 78	197.0 **Au** Gold 79	200.6 **Hg** Mercury 80	204.4 **Tl** Thallium 81	207.2 **Pb** Lead 82	209.0 **Bi** Bismuth 83	210.0 **Po** Polonium 84	210.0 **At** Astatine 85	222.0 **Rn** Radon 86
7	223.0 **Fr** Francium 87	226.0 **Ra** Radium 88	227.0 **Ac** † Actinium 89	261.1 **Rf** Rutherfordium 104	262.1 **Db** Dubnium 105	263.1 **Sg** Seaborgium 106	262.1 **Bh** Bohrium 107	— **Hs** Hassium 108	— **Mt** Meitnerium 109									

Revision Summary: Chemistry 3

Chemical reactions

- Scientists talk about 2 kinds of changes – chemical changes and physical changes.
- In a chemical change, one or more new substances are made. But there is no change in the total mass before and after the reaction. The same atoms are still there but they are combined in different ways. Chemical changes are usually difficult to reverse.
- In a physical change, no new substances are made. There is no change in mass. Physical changes are usually easy to reverse.
- Word equations show us the reactants and products in a reaction.
  e.g. magnesium + oxygen (reactants) ⟶ magnesium oxide (products)
  $2\,Mg + O_2 \longrightarrow 2MgO$
  iron + chlorine ⟶ iron chloride
  $2\,Fe + 3\,Cl_2 \longrightarrow 2\,FeCl_3$
  iron + sulphur ⟶ iron sulphide
  $Fe + S \longrightarrow FeS$

Acid rain has damaged these trees

- There are different ***types*** of reaction,
  e.g. when something gains oxygen, this is an **oxidation**.
- Some reactions are very useful to us,
  e.g. – setting superglue.
  – burning a fuel to keep us warm.
- Some chemical reactions are not useful to us,
  e.g. – iron rusting when it reacts with air and water.
  – fossil fuels making acid rain when they burn.
- Reactions which take in energy from the surroundings are **endothermic**.
  Reactions which give out energy are **exothermic**.

Caesium reacts violently with water

potassium
sodium
calcium
magnesium
zinc
iron
tin
copper

Reactivity Series

- Metals can react with:
  – oxygen to make oxides,
  e.g. copper + oxygen ⟶ copper oxide
  – water to make hydrogen gas,
  e.g. sodium + water ⟶ sodium hydroxide + hydrogen
  – acid to make hydrogen gas,
  e.g. magnesium + hydrochloric acid ⟶ magnesium chloride + hydrogen
  (Hydrogen gas 'pops' with a burning splint.)
- We can use these reactions to put metals in a league table of reactivity.
  This is called the Reactivity Series.
  e.g. magnesium is reactive – it is high up in the series.
  gold is unreactive – it is low in the series.
- A metal high in the Reactivity Series can displace one lower, from a solution of its salt.
  e.g. zinc displaces copper from copper sulphate solution
  zinc (silver-grey) + copper sulphate (blue solution) ⟶ zinc sulphate (colourless solution) + copper (pink/brown)
- The Reactivity Series can be used to make predictions about reactions.
  Q. magnesium + copper oxide ⟶ ?
  A. magnesium is more reactive than copper – so it displaces copper.
  magnesium (silver-grey) + copper oxide (black) ⟶ magnesium oxide (grey-white) + copper (brown)

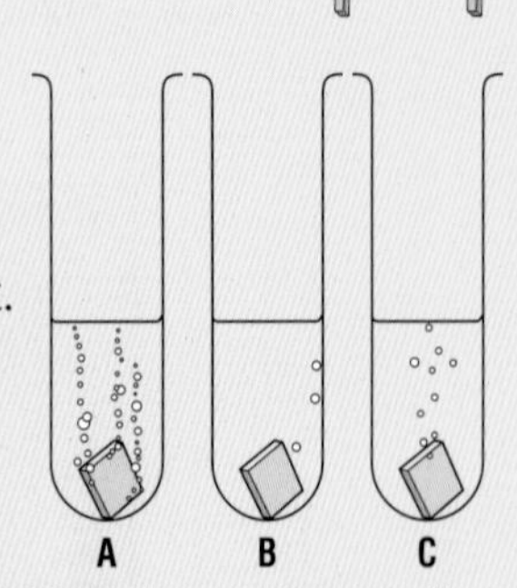

Metal A reacts fastest with acid. It is the most reactive. Metal B is the least reactive.

Revision Summary: Chemistry 4

Rocks and minerals

- Most rocks are mixtures. They contain different minerals.
- All rocks slowly crumble away. They are weakened by the weather. This process is called **weathering**. Weathering can be caused by water, wind and changes in temperature. Rocks expand as they get hotter and contract when they get cooler.
- The small pieces of rock rub against each other as they are moved, e.g. by wind and rivers. They wear away. This process is called **erosion**. The rock pieces can be transported and deposited in another place.
- There are 3 main types of rock – **igneous**, **sedimentary** and **metamorphic**. The 3 types form in different ways.

Conglomerate is a sedimentary rock

Slate is a metamorphic rock

Gabbro is an igneous rock

- Igneous rocks form when melted (molten) substances cool. They are hard and are made of crystals. Larger crystals form when the cooling is slow. Granite is an igneous rock.
- Sedimentary rocks form in layers. They are made when substances settle out in water. In hot weather, water from seas and lakes can run dry. A sediment can form in this way. Over time layers of sediment build up. Any water between the layers gets squeezed out. Minerals dissolved in the water get left behind. They 'cement' the sediments together. Sometimes these rocks contain fossils. The rocks are sometimes soft. Sandstone is a sedimentary rock.
- Metamorphic rocks are made when rocks are heated and/or squashed together. They form very slowly. They are usually very hard. Marble is a metamorphic rock.
- Over millions of years, one rock type can change into another. The rocks are recycled. We call this the **rock cycle**.

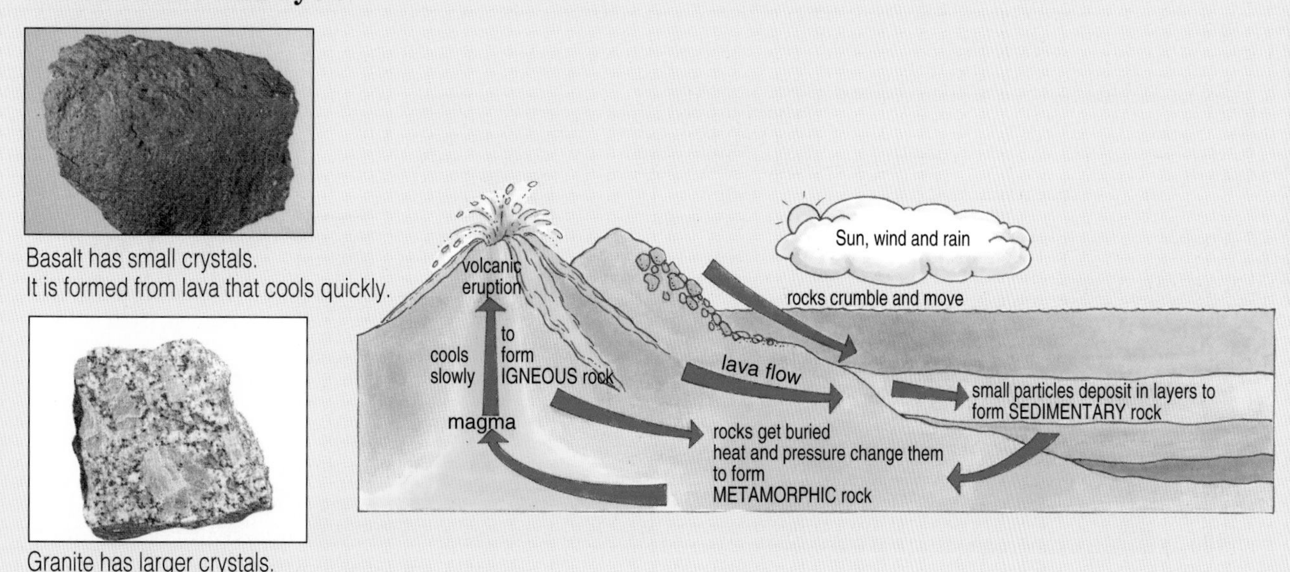

Basalt has small crystals.
It is formed from lava that cools quickly.

Granite has larger crystals.
It is formed from magma that cools slowly.

Revision Summary: Physics 1

Energy

- Energy can exist in several different forms, including: thermal energy (heat), light, sound, electrical, nuclear, chemical energy, etc.

  Movement energy is called kinetic energy.

  Gravitational potential energy is stored in an object that has been lifted up.
  Elastic (or 'strain') potential energy is stored in a stretched catapult.

- Fuels store energy. They give out the energy when they burn with oxygen.

  Coal, oil and natural gas are fossil fuels.
  Coal was formed over millions of years when dead plants were buried and squeezed by layers of rock.
  Oil and natural gas were made in a similar way from dead sea animals.

  These are ***non-renewable*** sources of energy (they get used up).
  Uranium is another non-renewable resource.

- We can also get energy from sunlight (solar energy), wind, waves, hydro-electric dams, and biomass (e.g. wood from trees). We can also get energy from tides and geothermal stations.
  These sources are ***renewable***.

- Most of our energy comes (indirectly) from the Sun. Solar heating causes winds, waves and rain.

- Electricity can be generated (made) from
  – non-renewable fuels (see the diagram below), or
  – renewable resources (e.g. a wind-generator).

- The laws of energy:
  1. The amount of energy before a transfer is always equal to the amount of energy after the transfer. The total amount of energy stays the same. We say the energy is 'conserved'.
  2. In energy transfers, the energy spreads out and becomes less useful to us.

- 1 kilojoule = 1000 joules.

- If the temperature of an object rises, its atoms vibrate more.
  The thermal energy (or heat) is the total energy of all the vibrating atoms in the object.

- To get a job done, energy must be transferred from one place to another.
  Energy can be transferred by electricity, by sound, by light, and by thermal (heat) transfer.

  Heat energy can be transferred by
  – conduction (through solids, liquids and gases),
  – convection (when liquids or gases move),
  – radiation (even through space).

Conduction from a hot object to a cold object:

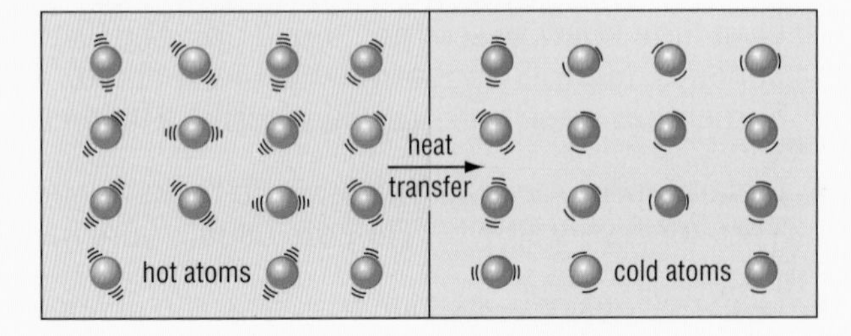

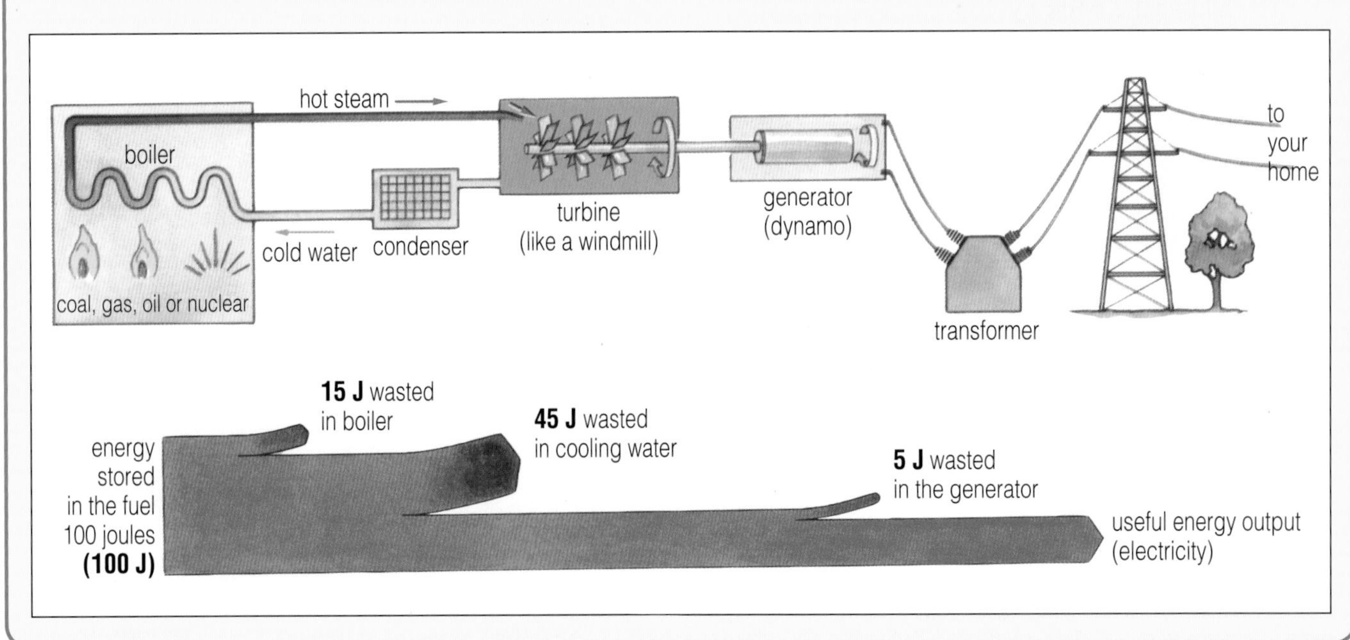

Revision Summary: Physics 2

Electricity

- For an electric current to flow, there must be a complete circuit.
- A conductor lets a current flow easily.
  A good conductor has a low resistance.
  An insulator does not let a current flow easily.
  It has a high resistance.
- Circuit diagrams are drawn using symbols:

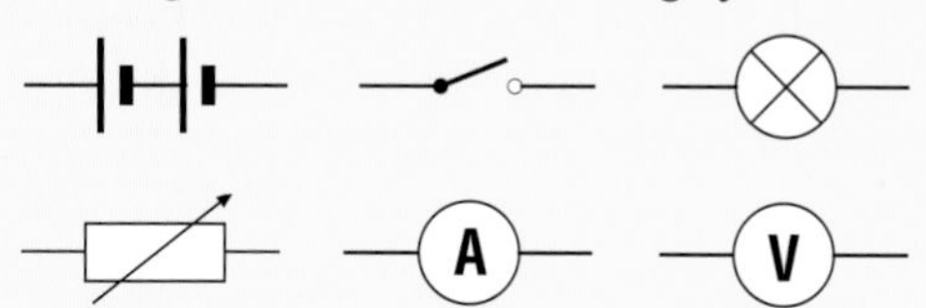

- An electric current is a flow of tiny electrons.
  It is measured in amperes (amps or A), using an ammeter.
  An ammeter must be connected in series in a circuit. It should have a low resistance.
- Voltage is measured by a voltmeter, placed across a component (in parallel).
  It should have a high resistance.
- In a series circuit, the same current goes through all the components:

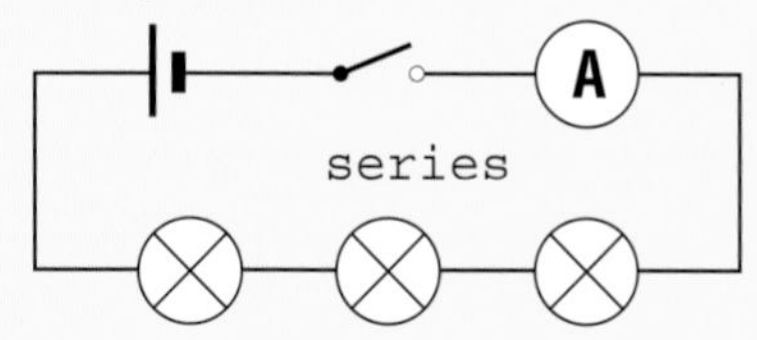

- To pass more current you can
  - add more cells pushing the same way,
  - reduce the number of bulbs in series.
- In a parallel circuit, there is more than one path.
  Some electrons go along one path and the rest go along the other path.

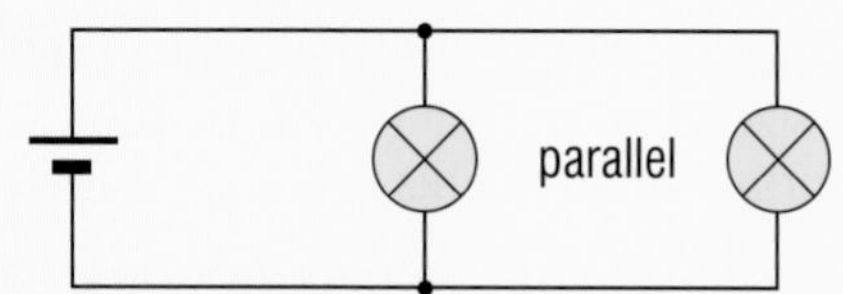

  To find the current in the main circuit, add up the currents in the branches.
- The electric current (flow of electrons) transfers energy from the cell to the bulbs.
  A bigger voltage means a bigger transfer of nergy.
  In the bulb the energy is transformed to heat and light energy.
- A fuse is used to protect a circuit. It melts if too much current flows, and breaks the circuit.

Magnetism

- The end of a compass that points North is called the North pole (N-pole) of the magnet.
- Two N-poles (or two S-poles) repel each other.
  An N-pole attracts an S-pole.
- Iron and steel can be magnetised. Other magnetic substances are nickel, cobalt and iron oxide.
- You can find the shape of the magnetic field round a bar magnet by using iron filings (or small plotting compasses).
  The shape is shown by 'lines of force' or 'field lines'.

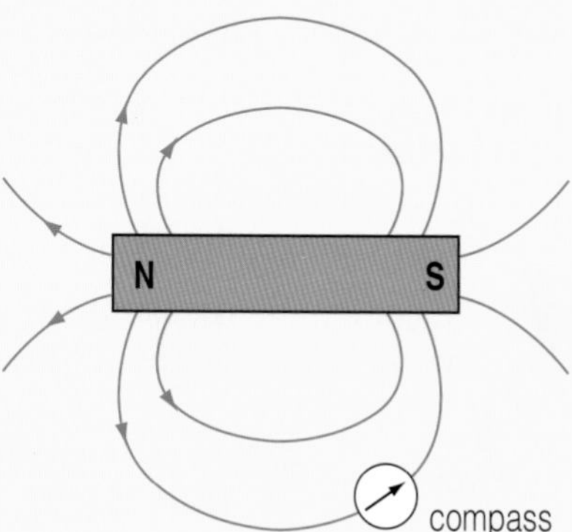

- The Earth has a magnetic field round it.
  A compass points along the field.
- A current in a coil produces a magnetic field like a bar magnet. It is an electromagnet.

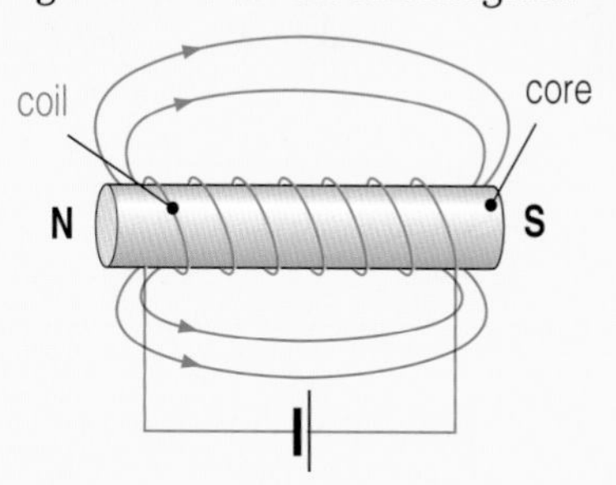

- The strength of an electromagnet can be increased by:
  (1) increasing the current in the coil,
  (2) increasing the number of turns on the coil, and
  (3) using an iron core.
  Reversing the current, reverses the North and South poles.
- In a relay, a small current in the coil of an electro-magnet is used to switch on a bigger current.
- In an electric bell, an electromagnet is used to attract an iron bar so that it breaks the circuit repeatedly. The vibrating iron bar makes the bell ring.

Revision Summary: Physics 3

Forces and motion

- Forces are pushes and pulls. They are measured in newtons (N).
  Friction is a force which always tries to slow down movement.
  Air resistance can be reduced by streamlining.
  Weight is a force. It is the pull of gravity by the Earth, downwards.
- If 2 forces on an object are equal and opposite, we say they are ***balanced*** forces. In this case there is no resultant force.
  When the forces are balanced, then the movement of the object is not changed (it stays still, or stays moving at a steady speed).
- If the forces on an object are not balanced, there is a ***resultant*** force.
  This resultant force makes the object speed up, or slow down, or change direction.
- The speed of a car can be measured in metres per second or kilometres per hour.

$$\textbf{Average speed} = \frac{\textbf{distance travelled}}{\textbf{time taken}}$$

- The motion of a car can be shown on a distance–time graph:

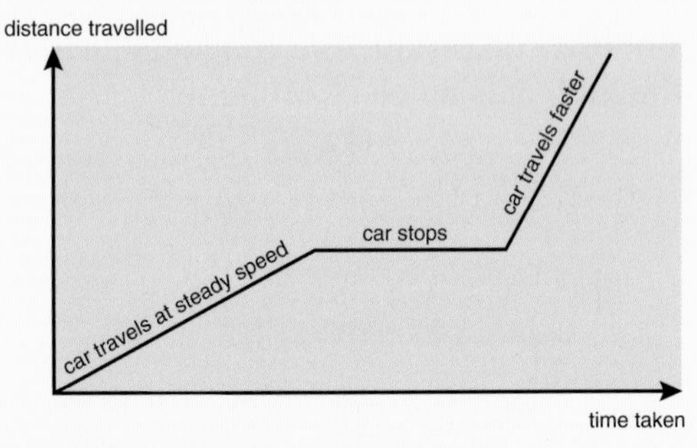

- Levers have many uses. A long spanner is easier to turn than a short spanner.
  The moment (or turning effect) of a force = force × distance of force from the pivot.

  The principle (or law) of moments says:

In equilibrium,
the anti-clockwise moments = the clockwise moments

- Pressure is measured in N/cm^2 or N/m^2 (also called a pascal, Pa).

$$\textbf{Pressure} = \frac{\textbf{force}}{\textbf{area}} \quad \frac{\text{(in newtons)}}{\text{(in cm}^2\text{ or m}^2\text{)}}$$

The Earth and beyond

- The Earth turns on its axis once in 24 hours. This is 1 day.
  This means that the Sun appears to rise in the East and set in the West. For the same reason, stars at night appear to move from East to West.
- The Earth travels round the Sun in an orbit, taking 1 year ($365\frac{1}{4}$ days).
  The Earth's axis is tilted (at $23\frac{1}{2}°$). This means that in summer the Sun is higher in the sky, and so day-time is longer and warmer than in winter.

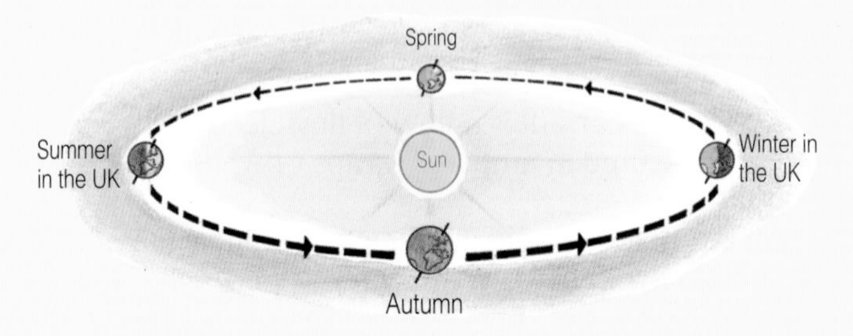

- The Moon moves round the Earth, taking 1 month for a complete orbit.
  The Moon shines because of sunlight on it.
  It shows different phases at different times of the month. If it goes into the shadow of the Earth, there is an eclipse of the Moon.
- The Sun is a star. It has 9 planets round it.

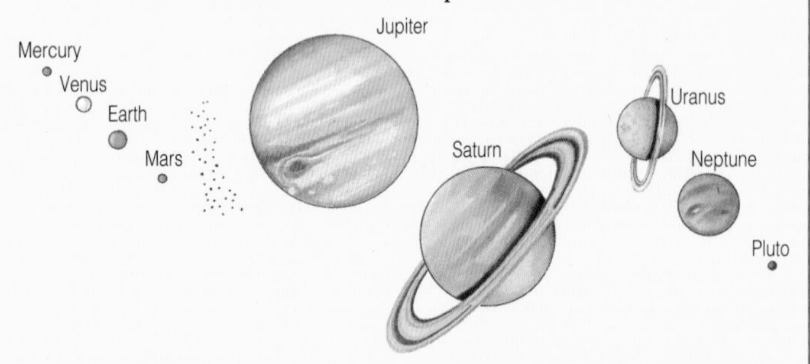

  The diagram above shows the order and the relative size of them (but they should be shown much further apart). This is the Solar System.
- A planet is held in orbit round the Sun by a gravitational force.
  The shape of each orbit is an ellipse.
- The Sun and other stars are very hot and make their own light.
  The planets and their moons shine only by reflecting the sunlight.
- Artificial satellites and space probes can be launched to observe the Earth and to explore the Solar System.

Revision Summary: Physics 4

Light

- In air, light travels in a straight line, at a very high speed. In glass it travels more slowly.
- Light travels much faster than sound.
- If light rays are stopped by an object, then a shadow is formed.
- You can see this page because some light rays (from a window or a lamp) are scattered by the paper, and then travel to your eye.
- The law of reflection: the angle of incidence (***i***) always equals the angle of reflection (***r***).
- The image in a plane (flat) mirror is as far behind the mirror as the object is in front.

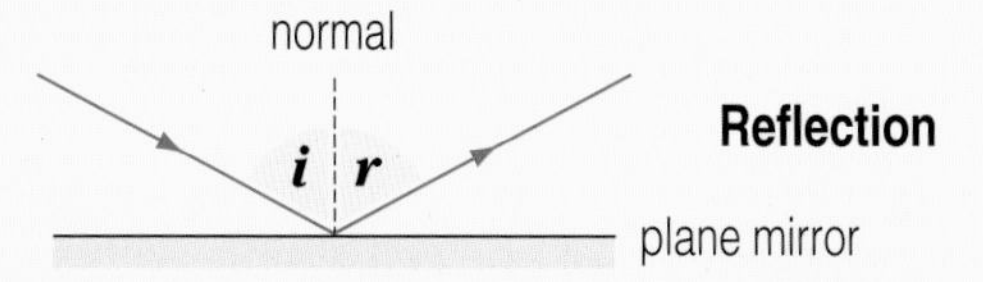

- Refraction: when a light ray goes into glass it is refracted (bent) *towards* the normal line. This is because the light slows down in the glass. When light comes out of glass, it is refracted *away from* the normal line.

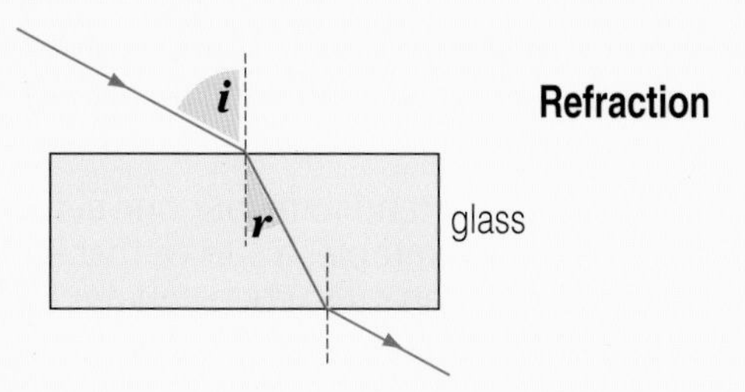

- When white light goes into a prism, it is dispersed into the 7 colours of the spectrum (ROY G BIV).

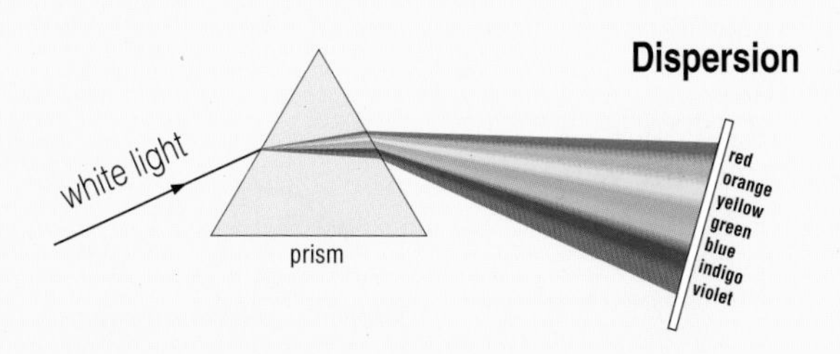

- A red filter will let through only red light (which we see). It absorbs all the other colours.
- A green object reflects green light (which we see). It absorbs all the other colours.

Sound

- Sound travels at about 330 metres per second (in air). Light travels much faster than this.
- Average speed $= \dfrac{\text{distance travelled}}{\text{time taken}}$
- All sounds are caused by vibrations.
- If a guitar string is vibrating, it sends out sound waves. These waves travel through the air to your ear. They transfer energy to your ear. The waves make your ear-drum vibrate and this sends messages to your brain.

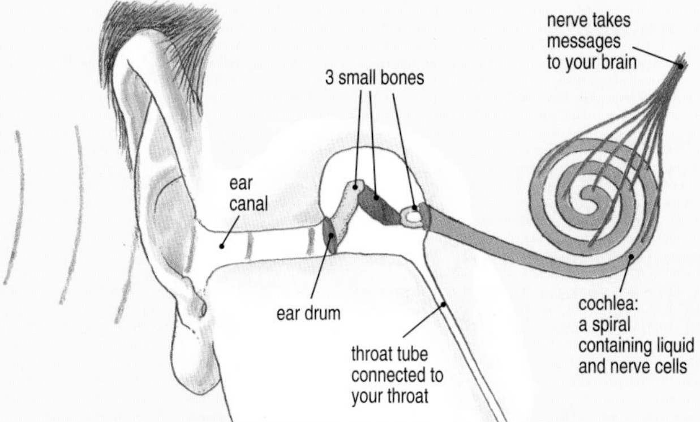

- Sound cannot travel through a vacuum. This is because there are no molecules to pass on the vibrations.
- To compare sound waves you can use a microphone and a CRO (oscilloscope).
- A loud sound has a large amplitude (a):

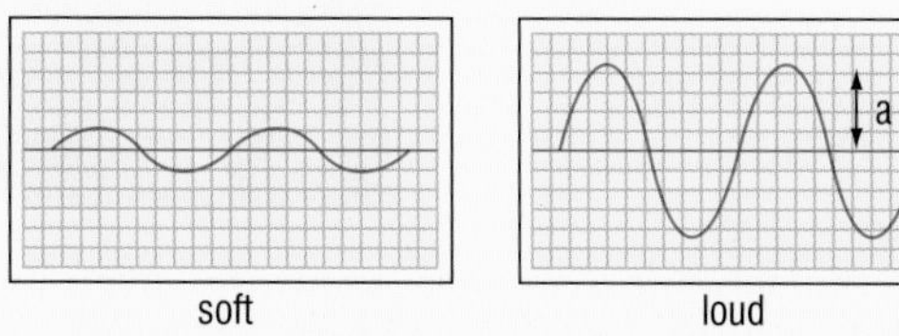

The loudness of a sound is measured in decibels (dB). Your ear is easily damaged by loud sounds.

- A high pitch sound has a high frequency:

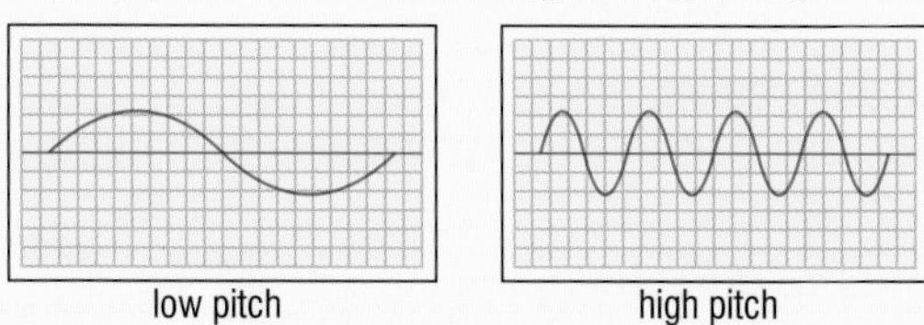

The frequency is measured in hertz (Hz). The range of frequencies that can be heard varies from person to person (from about 20 Hz up to about 20 000 Hz).

Glossary for Key Stage 3 Science

Abrasion
When a surface is worn away by rubbing.

Acceleration
The rate at which an object speeds up.

Acid
A sour substance which can attack metal, clothing or skin. The chemical opposite of an alkali.
When an acid is dissolved in water its solution has a pH number less than 7.

Adaptation
A feature that helps a plant or an animal to survive in changing conditions.

Adolescence
The time of change from a child to an adult, when both our bodies and our emotions change.

Aerobic respiration
The process that happens in cells whereby oxygen is used to release the energy from glucose.

Air resistance
A force, due to friction with the air, which pushes against a moving object, e.g. air resistance slows down a parachute.

Alkali
The chemical opposite of an acid. A base which dissolves in water. Its solution has a pH number more than 7.

Amphibian
An animal that lives on land and in water. It has moist skin and breeds in water.

Amplitude
The size of a vibration or wave, measured from its mid-point. A loud sound has a large amplitude.

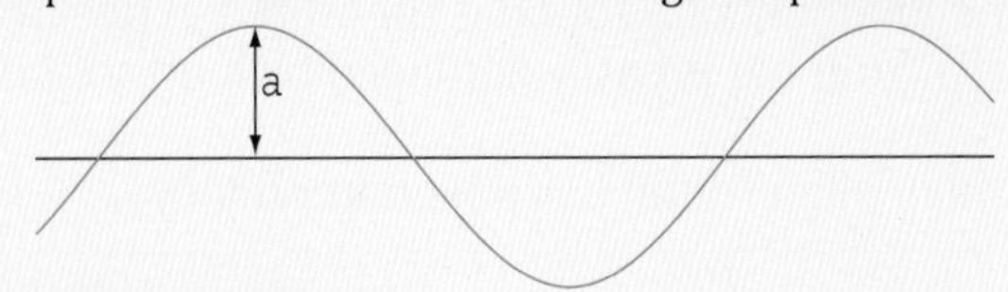

Anomalous result
A result that does not follow the general pattern in a set of data.

Antagonistic muscles
Muscles that work in pairs.
When one muscle contracts the other relaxes, e.g. your biceps and triceps.

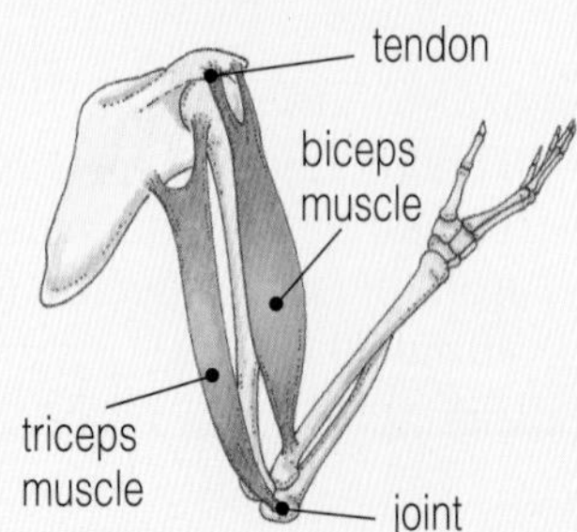

Antibiotic
A useful drug that helps your body fight a disease.

Artery
A blood vessel that carries blood away from the heart.

Asteroids
Small planets and pieces of rock hurtling through space. There is a belt of asteroids between Mars and Jupiter in the Solar System.

Atom
The smallest part of an element.

Axis
The Earth spins on its axis. It is an imaginary line passing through the Earth from the North Pole to the South Pole.

Bacteria
Microbes made up of one cell, visible with a microscope. Bacteria can grow quickly and some of them cause disease, e.g. pneumonia.

Balanced forces
Forces are balanced when they cancel out each other (see page 140). The object stays still, or continues to move at a steady speed in a straight line.

Base
The oxide, hydroxide or carbonate of a metal. (If a base dissolves in water it is called an alkali.)

Biased
Results that unfairly tend to favour one set of variables over another. For example, a survey that includes many more boys than girls when you are looking for a general pattern.

Biomass fuel
Fuel (e.g. wood) made from growing plants.

Braking distance
The distance a car travels *after* the brake is pressed.

Bronchus
One of the tubes at the bottom of the wind-pipe (trachea) that lead to the lungs.

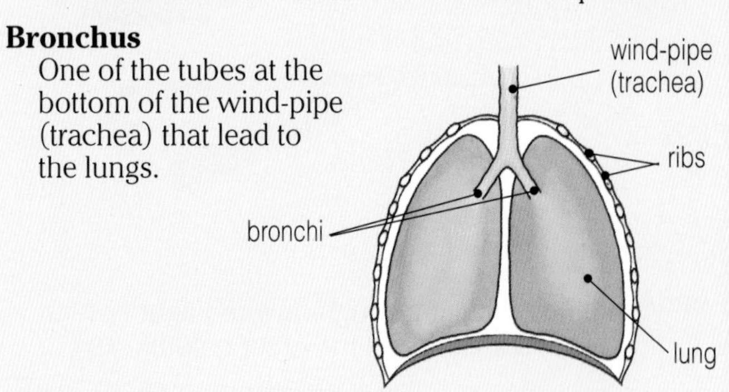

Capillaries
Tiny blood vessels that let substances like oxygen, food and carbon dioxide pass into and out of the blood.

Carbohydrate
Your body's fuel. Food like glucose that gives you your energy.

Carnivores
Animals that eat only other animals – meat-eaters.

Catalytic converter
Device fitted to a car's exhaust system to reduce the pollutant gases given out.

Caustic
A caustic substance is corrosive.

Cell membrane
The structure that surrounds a cell and controls what goes in and out.

Cell wall
The strong layer on the outside of a plant cell that supports the cell.

Cells
The 'building blocks' of life, made up of a *cell membrane*, *cytoplasm* and *nucleus*.

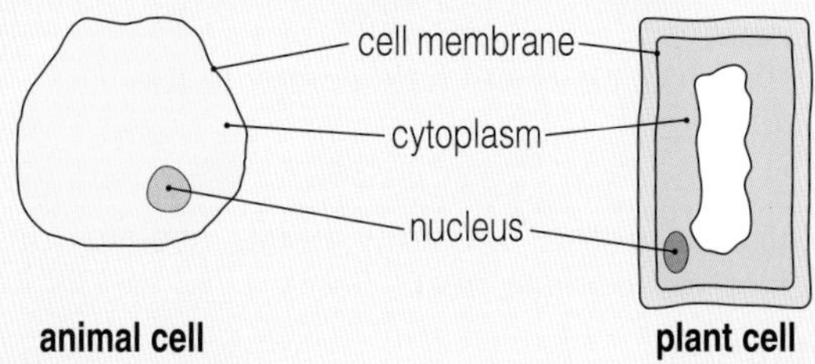

Chemical change
A change which makes a new substance, e.g. coal burning.

Chemical energy
The energy stored in substances, e.g. foods and fuels are useful stores of chemical energy.

Chlorophyll
A green chemical in plants used to trap light energy for photosynthesis.

Chloroplasts
Tiny, round structures found inside plant cells. They capture light energy and use it to make food in photosynthesis.

Chromatography
A method used to separate mixtures of substances, usually coloured ones.

Cilia
Very small threads or hairs on the surface of some cells.

Classify
To sort things out into different groups or sets.

Clone
Genetically-identical living things.

Combustion
The reaction which occurs when a substance burns in oxygen, giving out heat energy.

Community
The group of animals and plants that we find in a particular habitat.

Competition
A struggle for survival. Living things compete for scarce resources, e.g. space.

Component
One of the parts that make up an electric circuit, e.g. battery, switch, bulb.

Composition
The type and amount of each element in a compound.

Compound
A substance made when 2 or more elements are chemically joined together, e.g. water is a compound made from hydrogen and oxygen.

Conductor
An electrical conductor allows a current to flow through it. A thermal conductor allows heat energy to pass through it. All metals are good conductors.

Control variables
The factors we must keep the same during an investigation to make sure we carry out a fair test.

Convection
The transfer of heat by currents in a liquid or a gas.

Correlation
The strength of the link or connection between two variables being investigated.

Corrosive
A corrosive substance can eat away another substance by attacking it chemically.

Cytoplasm
The jelly-like part of the cell where many chemical reactions take place.

Dependent variable (or outcome variable)
When you do a fair test, this is the factor that you measure or observe in each test, in order to see the effect of varying another factor.
For example, if you investigate 'How does temperature affect the time taken for sugar to dissolve?', then the dependent variable is the time taken.

Diffusion
The process of particles moving and mixing of their own accord, without being stirred or shaken.

Digestion
Breaking down food so that it is small enough to pass through the gut into the blood.

Dispersion
The splitting of a beam of white light into the 7 colours of the spectrum, by passing it through a prism.

Displacement
When one element takes the place of another in a compound. For example,

magnesium + copper sulphate → magnesium sulphate + copper

This is called a displacement reaction.

Dissipation of energy
When energy is spread out by being transferred to lots of different places.

Distillation
A way to separate a liquid from a mixture of liquids, by boiling off the substances at different temperatures.

Dormant
Inactive, e.g. a dormant volcano is one which has not erupted for a long time.

Drag
Friction caused by an object travelling through a liquid or gas. For example, friction caused by air resistance.

Drug
A chemical that alters the way in which your body works, e.g. alcohol, cannabis, nicotine, solvents.

Dynamo
A machine that transfers kinetic energy to electrical energy.

Eclipse
A ***lunar eclipse*** is when the shadow of the Earth falls on the Moon.

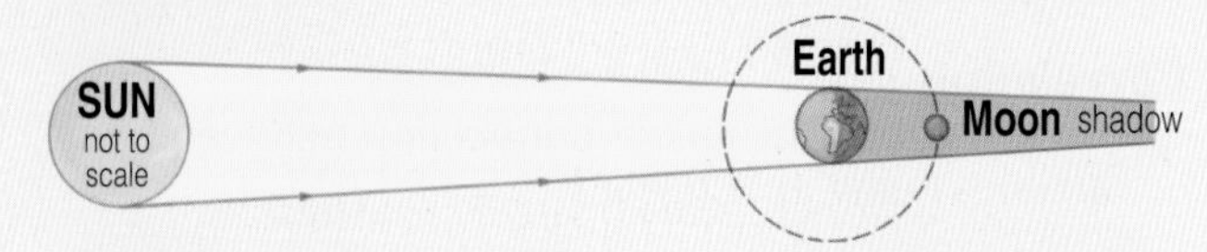

A ***solar eclipse*** is when the Sun is blotted out (totally or partially) by the Moon.

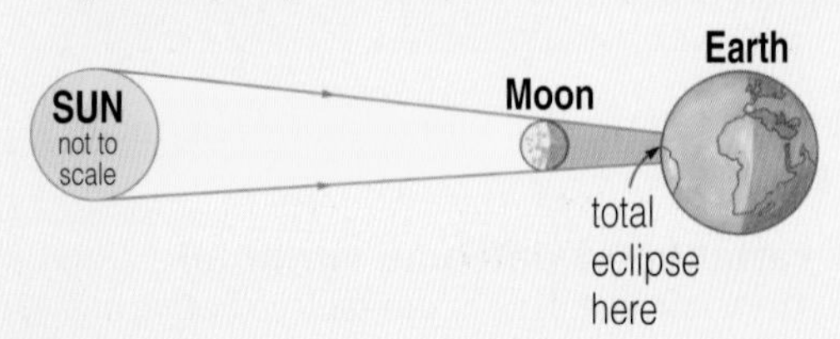

Egg
Female sex cell.

Electric current
A flow of electric charges (electrons).
It is measured in amps (A) by an ammeter.

Electro-magnet
A coil of wire becomes a magnet when a current flows through it.

Electron
A tiny particle with a negative charge.

Element
A substance that is made of only one type of atom.

Embryo
A fertilised egg grows into an embryo and eventually into a baby.

Endothermic
A reaction that *takes in* heat energy from the surroundings.

Energy transfer
See *Transfer of energy*.

Enzymes
Chemicals that act like catalysts to speed up digestion of our food.

Equation
A shorthand way of showing the changes that take place in a chemical reaction

e.g. iron + sulphur → iron sulphide
$Fe + S \rightarrow FeS$

Equilibrium
A balanced situation, when all the forces cancel out each other.

Erosion
The wearing away of rocks.

Evaluate
a) Your method,
to judge how effective the method you used was in collecting reliable data.
b) Your conclusion,
to judge how strong the evidence is that you have used to draw a conclusion.

Exothermic
A reaction that *gives out* heat energy to the surroundings.

Fat
Food used as a store of energy and to insulate our bodies so we lose less heat.

Fermentation
The reaction when sugar is turned into alcohol.

Fertilisation
When sex cells join together to make a new individual, e.g. a sperm and an egg, or a pollen grain nucleus and an ovule nucleus.

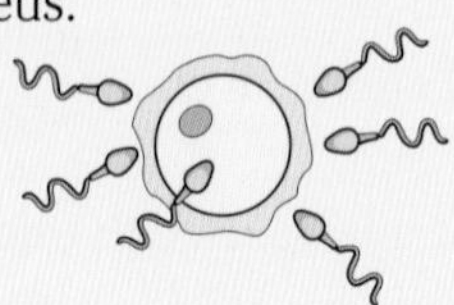

Fertilisers
The nutrients that can be added to the soil if they are in short supply.

Fetus
An embryo that has developed its main features, e.g. in humans after about 3 months.

Fibre
Food that we get from plants that cannot be digested. It gives the gut muscles something to push against.

Filtration
A process used to separate undissolved solids from liquids.

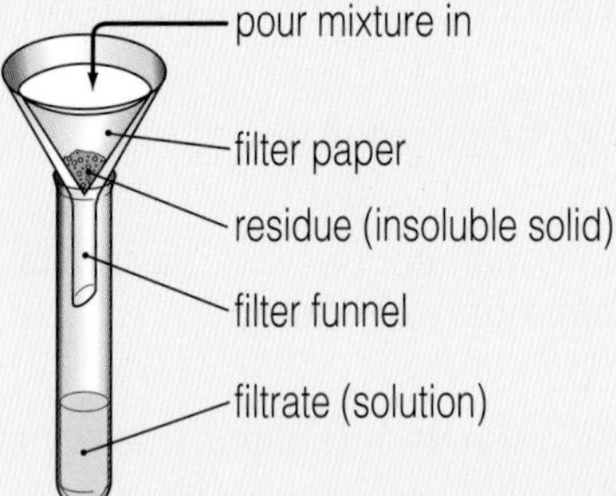

Flowers
The organs that many plants use to reproduce by making seeds.

Food chain
A diagram that shows how food energy is passed between plants and animals.

Food web
A diagram that shows a number of food chains linked together.

Formula
A combination of symbols to show the elements which a compound contains,
e.g. MgO is the formula for magnesium oxide.

Fossil
The remains of an animal or plant which have been preserved in rocks.

Fossil fuels
A fuel made from the remains of plants and animals that died millions of years ago, e.g. coal, oil, natural gas.

Frequency
The number of complete vibrations in each second.
A sound with a high frequency has a high pitch.

Friction
A force when 2 surfaces rub together. It always pushes against the movement.

Fuel
A substance that is burned in air (oxygen) to give out energy.

Fungi
Moulds, such as yeast or mushrooms, that produce spores.

Fungicide
A chemical which kills fungi that attack crops.

Fuse
A safety device in an electrical circuit. It is a piece of wire that heats up and melts, breaking the circuit, if too much current passes through it.

Gamete
The male or female reproductive cells.

Gas
A substance which is light, has the shape of its container and is easily squashed. The particles in a gas are far apart. They move quickly and in all directions.

Genes
Found in chromosomes, they control the inherited features of living things.

Gestation
The time from fertilisation to birth, e.g. in humans the gestation period is about 40 weeks.

Global warming
The build up of 'greenhouse' gases that is causing the temperature of the Earth to increase.

Gravity, gravitational force
A force of attraction between 2 objects.
The pull of gravity on you is your weight.

Group
All the elements in one column down the periodic table.

Habitat
The place where a plant or animal lives.

Haemoglobin
The substance in red blood cells that transports oxygen around the body.

Herbicide
A chemical used to kill weeds that are in competition with a crop.

Herbivores
Animals that eat only plants.

Hibernate
To remain inactive throughout the winter months.

Igneous rock
A rock formed by molten (melted) material cooling down.

Image
When you look in a mirror, you see an image of yourself.

Immune
Not being able to catch a particular disease because you have the antibodies in your blood to fight it.

Inherited
The features that are passed on from parents to their offspring.

Independent variable (or **input variable)**
The factor that you choose to change in an investigation.
For example, if you investigate 'How does temperature affect the time taken for sugar to dissolve**?**', then the independent variable is the temperature.

Indicator
A substance that changes colour depending on the pH of the solution you add it to.

Insecticide
A chemical that kills insects that feed on crops.

Insulator
An electrical insulator does not allow a current to flow easily. A thermal insulator does not let heat energy flow easily.

Intestine
Tube below the stomach where food is digested and absorbed.

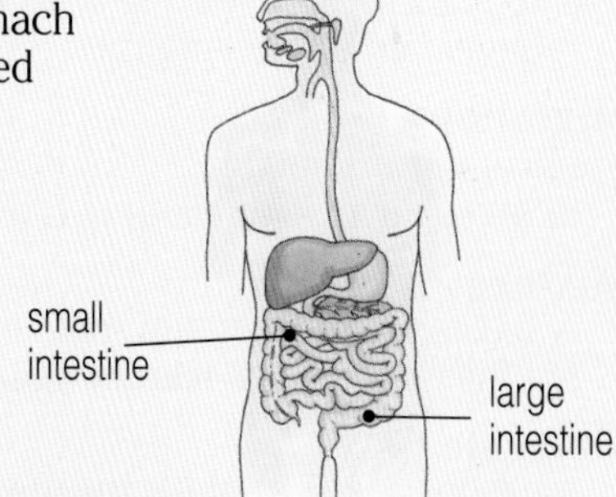

Invertebrate
An animal without a backbone.

Joint
The point where 2 bones meet. Joints usually allow movement.

Kinetic energy
The energy of something which is moving.

Law of reflection
When light rays bounce off a mirror:

angle of incidence = angle of reflection.

Lava
Molten rock ejected from a volcano.

Lever
A simple machine that produces a bigger force or movement than we apply.

Liquid
A substance which has the shape of its container, can be poured and is not easily squashed. The particles in a liquid are quite close together but free to move.

Lungs
The organs in our body that collect oxygen and get rid of carbon dioxide.

Magma
Hot molten rock below the Earth's surface.

Magnetic field
The area round a magnet where it attracts or repels another magnet.

Mammal
Warm-blooded animals with fur or hair that suckle their young.

Menstruation
The discharge of blood and lining of the uterus from the vagina. This happens at the end of each menstrual cycle in which an egg has not been fertilised.

Metal
An element which is a good conductor and is usually shiny, e.g. copper.

Metamorphic rock
A rock formed by heating and compressing (squeezing) an existing rock.

Migration
Moving from one place to another in different seasons to avoid adverse or harsh conditions.

Mixture
A substance made when some elements or compounds are mixed together. It is *not* a pure substance.

Molecule
A group of atoms joined together.

Moment
The turning effect of a force.

Moment = force × distance from the pivot.

Muscle
Structures that contract and relax to move bones at joints.

Neutral
Something which is neither an acid nor an alkali.

Neutralisation
The chemical reaction of an acid with a base, in which they cancel each other out.

Non-metal
An element which does not conduct electricity. (The exception to this is graphite – a form of carbon which is a non-metal, but it does conduct).

Non-renewable resources
Energy sources that are used up and not replaced, e.g. fossil fuels.

Nucleus of a cell
A round structure that controls the cell and contains the instructions to make more cells.

Nutrients
The chemicals needed by plants for healthy growth, e.g. nitrates, phosphates.

Opaque
An opaque object will not let light pass through it.

Orbit
The path of a planet or a satellite.
Its shape is usually an ellipse (oval).

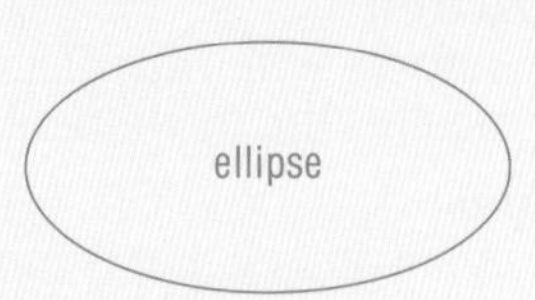

Organ
A structure made up of different tissues that work together to do a particular job.

Ovary
Where the eggs are made in a female.

Oviduct
A tube that carries an egg from the ovary to the uterus.

Oxidation
The reaction when oxygen is added to a substance.

Ozone depletion
The destruction of the ozone layer in our atmosphere. Ozone protects us from the Sun's harmful ultra-violet radiation.

Palisade cells
Cells in which most photosynthesis takes place. They are found in the upper part of a leaf.

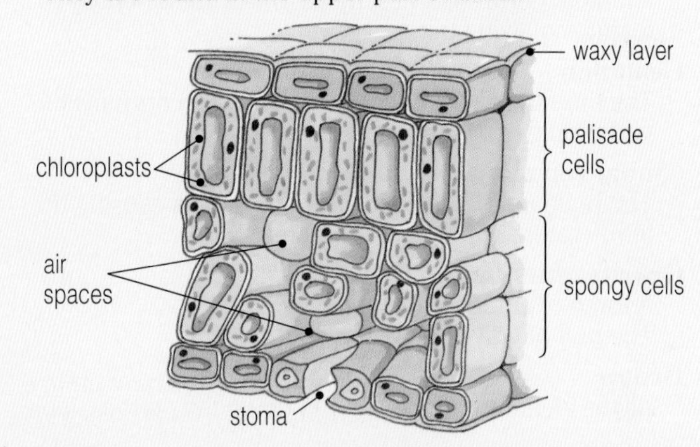

Parallel circuit
A way of connecting things in an electric circuit, so that the current divides and passes through different branches.

Period (1)
When the lining of the uterus breaks down and blood and cells leave the body through the vagina.

Period (2)
All the elements in one row across the periodic table.

Periodic table
An arrangement of elements forming groups and periods.

Pesticide
A chemical that kills insects, weeds or fungi that damage crops.

pH number
A number which shows how strong an acid or alkali is.
Acids have pH below 7 (0–6).
Alkalis have pH above 7 (8–14).

Photosynthesis
The process by which green plants use light energy to turn carbon dioxide and water into sugars:

$$\text{carbon dioxide} + \text{water} \xrightarrow[\text{chlorophyll}]{\text{light and}} \text{sugar} + \text{oxygen}$$

Physical change
A change in which no new substance is made.
The substance just changes to a different state, e.g. water boiling.

Pitch
A whistle has a high pitch, a bass guitar has a low pitch.

Placenta
A structure that forms in the uterus allowing the blood of the baby and the blood of the mother to come close together.

Pollination
The transfer of pollen from the anthers to the stigma of a flower.

Population
A group of animals or plants of the same species living in the same habitat.

Porosity
The ability to absorb a liquid such as water, e.g. sandstone is a porous rock.

Potential energy
Stored energy, e.g. a bike at the top of a hill has gravitational potential energy.

Predator
An animal that hunts and eats other animals.

Prediction
A statement that describes and explains what you think will happen in an investigation.

Pressure
A large force pressing on a small area gives a high pressure.

$$\text{Pressure} = \frac{\text{force}}{\text{area}}$$

Prey
An animal that is eaten by a predator, e.g. a rabbit is prey for the fox.

Principle of conservation of energy
The amount of energy before a transfer is always equal to the amount of energy after the transfer. The energy is 'conserved'.

Producers
Green plants that make their own food by photosynthesis.

Product
A substance made as a result of a chemical reaction.

Proportional
The link between 2 variables, e.g. the extension of a spring is directly proportional to the load on it, so if you double the load, the extension is also doubled.

Protein
Food needed for the growth and repair of cells.

Puberty
The age at which the sexual organs become developed.

Pumice
A light, porous rock formed from lava.

Pyramid of numbers
A diagram to show how many living things there are at each level in a food chain.

Qualitative observations
Descriptive evidence from an investigation, e.g. your observations of different pieces of material left buried for several weeks to see which are biodegradable.

Quantitative data
Numerical evidence produced from your measurements during an investigation.

Radiation
Rays of light, X-rays, radio waves, etc., including the transfer of energy through a vacuum.

Reaction
A chemical change which makes a new substance.

Reactivity series
A list of elements in order of their reactivity. The most reactive element is put at the top of the list.

Reduction
A reaction when oxygen is removed, e.g. copper oxide is *reduced* to copper.

Reflection
When light bounces off an object.
Angle i = angle r

Refraction
A ray of light passing from one substance into another is bent (refracted).

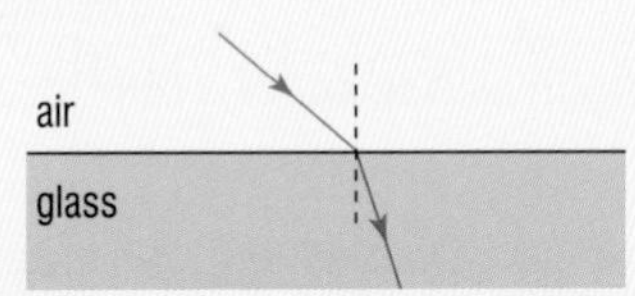

Relay
A switch that is operated by an electro-magnet.
A small current can switch on a large current.

Reliability
A measure of the trust you can put in your results.
For example, if you were to collect your data again would you get the same results?
In some investigations it is difficult to get precise measurements, e.g. measuring how high a ball bounces. You can improve the reliability of your results by repeating readings and taking averages.

Renewable energy resources
Energy sources that do not get used up,
e.g. solar energy, wind, waves, tides, etc.

Reptile
An animal with dry, scaly skin that lays eggs with soft shells.

Resistance
A thin wire gives more resistance to an electric current than a thick wire.

Respiration
The release of energy from food in our cells. Usually using up oxygen and producing carbon dioxide.

glucose + oxygen → carbon dioxide + water + energy

Resultant force
The result of *unbalanced forces*.

Rock cycle
A cycle that means that one type of rock can be changed into another type of rock over a period of time.

Salt
A substance made when an acid and a base react together.

Sample size
The number of subjects included in an enquiry.
For example, in biological investigations it is difficult to control all the variables. So you can improve your evidence by collecting data from more plants, animals or places. This is called increasing your sample size.

Satellite
An object that goes round a planet or a star,
e.g. the Moon goes round the Earth.

Saturated solution
A solution in which no more solute can dissolve at that temperature.

Scattering
When rays of light hit a rough surface (like paper) they reflect off in all directions.

Sedimentary rock
A rock formed by squashing together layers of material that settle out in water.

Selective breeding
Choosing which animals and plants to breed in order to pass on useful features to the offspring, e.g. high milk yield.

Series circuit
A way of connecting things in an electric circuit,
so that the current flows through each one in turn.

Solar System
The Sun and the 9 planets that go round it:
Mercury, Venus, Earth, Mars, (asteroids), Jupiter,
Saturn, Uranus, Neptune, Pluto.

Solid
A substance which has a fixed shape, is not runny and is not squashed easily. The particles in a solid are packed very closely together – they vibrate but do not move from place to place.

Soluble
Describes something which dissolves,
e.g. salt is soluble in water.

Solute
The solid that dissolves to make a solution.

Solution
The clear liquid made when a solute dissolves in a solvent, e.g. salt (solute) dissolves in water (solvent) to make salt solution.

Solvent
The liquid that dissolves the solute to make a solution.

Species
A type of living thing that breeds and produces fertile offspring.

Spectrum
The colours of the rainbow that can be separated when white light is passed through a prism:
red, orange, yellow, green, blue, indigo, violet (ROY G. BIV).

Speed
How fast an object is moving.

$$\text{Speed} = \frac{\text{distance travelled}}{\text{time taken}}$$

Sperm
Male sex cell.

States of matter
The 3 states in which matter can be found: *solid*, *liquid* and *gas*.

Temperature
How hot or cold something is.
It is measured in °C, using a thermometer.

Testis
Where the sperms are made in a male.

Thermal energy
Another name for heat energy.

Thermal transfer
When a cup of tea cools down, there is a transfer of thermal energy (heat) from the cup to the surroundings. This transfer can be by conduction, convection, radiation and evaporation.

Thinking distance
The distance travelled in a car during the driver's reaction time.

Tissue
A group of similar cells that look the same and do the same job.

Toxins
Poisons produced from bacteria and other microbes.

Trachea
The wind-pipe taking air to and from the lungs.

Transfer of energy
The movement of energy from one place to another, for a job to be done.

Transformation of energy
When energy changes from one form to another, e.g. when paper burns, chemical energy is changed to heat and light energy.

Unbalanced forces
If 2 forces do not cancel out each other, they are unbalanced. There will be a resultant force. The object will change its speed or change its direction.

Universal indicator
A liquid which changes colour when acids or alkalis are added to it. It shows whether the acid or alkali is strong or weak.

Upthrust
Upward force produced on an object in a liquid or a gas. There is a very small upthrust in a gas.

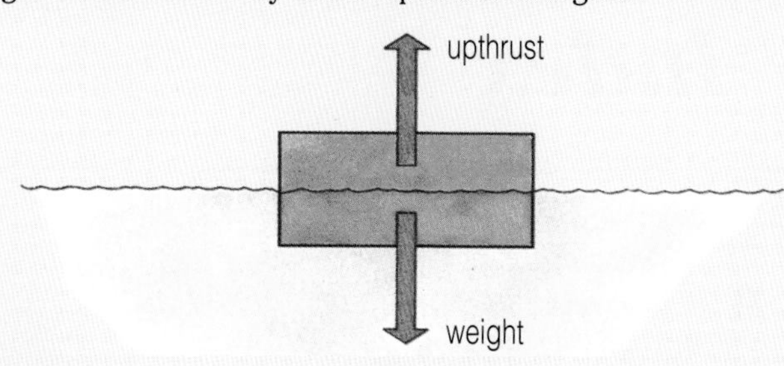

Uterus
The womb, where a fertilised egg settles and grows into a baby.

Vaccination
Protection against a disease by introducing into the body a harmless sample of the microbe that causes infection.

Vacuole
The space in a plant cell that is filled with a watery solution called cell sap.

Validity of conclusions
A measure of the certainty you can put on your conclusions drawn from the data that you collected in your investigation.

Variable
The things (factors) that can change (or vary) in an investigation.

Variation
Differences between *different* species, e.g. between dogs and cats, or between individuals of the *same* species, e.g. people in your class.

Vein
A blood vessel that carries blood back to the heart.

Vertebrate
An animal that has a backbone.

Vibrating
Moving backwards and forwards quickly, e.g. the particles in a solid vibrate.

Viruses
Extremely small microbes which are not visible with a microscope. Many viruses spread disease by invading cells and copying themselves, e.g. influenza.

Vitamins
Complex chemicals needed in small amounts to keep us healthy, e.g. vitamin C.

Wavelength
The distance between 2 peaks of a wave.

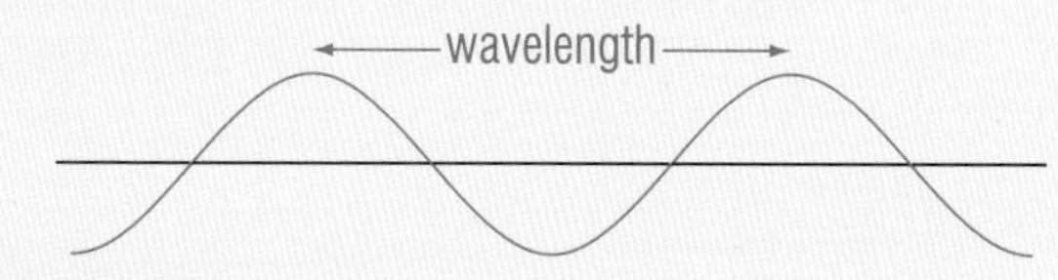

Weathering
The crumbling away of rocks caused by weather conditions such as wind and rain.

Index